Innovation in Energy Systems - New Technologies for Changing Paradigms

Edited by Taha Selim Ustun

Published in London, United Kingdom

IntechOpen

Supporting open minds since 2005

Innovation in Energy Systems - New Technologies for Changing Paradigms
http://dx.doi.org/10.5772/intechopen.81310
Edited by Taha Selim Ustun

Contributors
Vikas Khare, Aaquil Bunglowala, Vítor Monteiro, Jose Afonso, Tiago Sousa, Luiz Cardoso, José Gabriel Pinto, Joao Luiz Afonso, Salem Elsaiah, Thomas Caravella, Christopher Austell, Christian Alvarez, Grazyna Jastrzebska, Inderpreet Kaur, Harpreet Kaur, M Edwin, S Joseph Sekhar, M Saranya Nair, Mario Jorizzo, Nino Di Franco, Omar H. Abdalla, Azza Aly, Tuğçe Demirdelen, Kemal Aygul, Burak Esenboga, Abdurrahman Yavuzdeğer, Firat Ekinci, Mehmet Tumay

Notice
Statements and opinions expressed in the chapters are these of the individual contributors and not necessarily those of the editors or publisher. No responsibility is accepted for the accuracy of information contained in the published chapters. The publisher assumes no responsibility for any damage or injury to persons or property arising out of the use of any materials, instructions, methods or ideas contained in the book.

First published in London, United Kingdom, 2019 by IntechOpen
IntechOpen is the global imprint of INTECHOPEN LIMITED, registered in England and Wales, registration number: 11086078, 7th floor, 10 Lower Thames Street, London, EC3R 6AF, United Kingdom
Printed in Croatia

British Library Cataloguing-in-Publication Data
A catalogue record for this book is available from the British Library

Additional hard and PDF copies can be obtained from orders@intechopen.com

Innovation in Energy Systems - New Technologies for Changing Paradigms
Edited by Taha Selim Ustun
p. cm.
Print ISBN 978-1-78984-107-7
Online ISBN 978-1-78984-108-4
eBook (PDF) ISBN 978-1-78985-591-3

An electronic version of this book is freely available, thanks to the support of libraries working with Knowledge Unlatched. KU is a collaborative initiative designed to make high quality books Open Access for the public good. More information about the initiative and links to the Open Access version can be found at www.knowledgeunlatched.org

For EU product safety concerns:
IN TECH d.o.o., Prolaz Marije Krucifikse Kozulić 3, 51000 Rijeka, Croatia, info@intechopen.com or visit our website at intechopen.com.

Meet the editor

Taha Selim Ustun received his Ph.D. degree in electrical engineering from Victoria University, Melbourne, VIC, Australia. He has been an Assistant Professor of electrical engineering with ECE, Carnegie Mellon University, Pittsburgh, PA, USA. He is currently a Researcher with the Fukushima Renewable Energy Institute, AIST (FREA), where he leads the Smart Grid Cybersecurity Laboratory. He has edited several books and special issues with international publishing houses. His current research interests include power system protection, communication in power networks, distributed generation, microgrids, electric vehicle integration, and cybersecurity in smart grids. He has delivered several invited talks for different conferences and organizations such as the Qatar Foundation, the World Energy Council, the Waterloo Global Science Initiative, and the European Union Energy Initiative (EUEI).

Contents

Preface **XI**

Section 1
System Planning **1**

Chapter 1 **3**
Solar-Wind Energy Assessment by Big Data Analysis
by Vikas Khare and Aaquil Bunglowala

Chapter 2 **25**
Technical Requirements for Connecting Solar Power Plants to Electricity Networks
by Omar H. Abdalla and Azza A.A. Mostafa

Chapter 3 **51**
Energy Return on Investment Analysis of a Solar Photovoltaic System
by Harpreet Kaur and Inderpreet Kaur

Chapter 4 **93**
Efficiency, Energy Saving, and Rational Use of Energy: Different Terms for Different Policies
by Nino Di Franco and Mario Jorizzo

Section 2
Microgrids **113**

Chapter 5 **115**
Use of Micro-Cogeneration in Microgrids to Support Renewables
by Kemal Aygul, Burak Esenboga, Abdurrahman Yavuzdeger, Fırat Ekinci, Tugce Demirdelen and Mehmet Tumay

Chapter 6 **139**
Hybrid Maritime Microgrids: A Quest for Future Onboard Integrated Marine Power Systems
by Thomas Caravella, Christopher Austell, Christian Brady-Alvarez and Salem Elsaiah

Section 3
Electric Vehicles **153**

Chapter 7 **155**
Vehicle Electrification: Technologies, Challenges, and a Global Perspective for Smart Grids
by Vitor Monteiro, Jose A. Afonso,Tiago J.C. Sousa, Luiz L. Cardoso, Jose Gabriel Pinto and Joao L. Afonso

Section 4
Industrial Applications **205**

Chapter 8 **207**
The Innovative Gaildorf Wind-Water Project Guarantees Reliability of Power Supply
by Grażyna Frydrychowicz-Jastrzębska

Chapter 9 **225**
Hybrid Energy-Based Chilling System for Food Preservation in Remote Areas
by Edwin Mohan, Saranya Nair Mohan and Joseph Sekhar Santhappan

Preface

It has been a little over a century since the inception of interconnected networks and little has changed in the way that they are operated. Demand-supply balance methods, protection schemes, business models for electric power companies, and future development considerations have remained the same until very recently. Distributed generators, storage devices, and electric vehicles have become widespread and disrupted century-old bulk generation – bulk transmission operation. Distribution networks are no longer passive networks and now contribute to power generation. Old billing and energy trading schemes cannot accommodate this change and need revision. Furthermore, bidirectional power flow is an unprecedented phenomenon in distribution networks and traditional protection schemes require a thorough fix for proper operation.

Most distributed generators are based on intermittent resources and may not be able to supply energy all of the time. This requires deterministic demand-supply balance methods to be amended and consider probabilistic generation contribution. Also, most of the distributed generators are inverter-interfaced, in contrast to bulky rotating machines used in power networks. Inverters are based on power electronics and exhibit very different behavior under disturbances such as voltage swings and frequency variations.

Operational rules such as droop control have to be updated to include these new generation generators. In addition to these technical changes, business models and development considerations have changed drastically. Solar home systems, rural stand-alone microgrids, and DC microgrids have proliferated around the world as new business models. Power companies are looking at ways to include consumers in their operation decisions through schemes such as demand side management and vehicle to grid support.

All of these mean only one thing. It is a very exciting time for researchers working in the energy field. There are lots of new ideas that are proposed for all these different aspects. More intelligent devices are developed, new algorithms are implemented for better optimization while new start-up companies appear every day with ground-breaking business models. Traditional power systems are analog, unilateral, and mostly passive. New age power networks are smart, digital, incorporate communication for better operation, and involve collaborative decision-making. This book aims to cover new technologies, methods, and approaches developed to meet the needs of this changing field.

Taha Selim Ustun (PhD)
Fukushima Renewable Energy Institute, AIST (FREA),
Japan

Section 1

System Planning

Chapter 1

Solar-Wind Energy Assessment by Big Data Analysis

Vikas Khare and Aaquil Bunglowala

Abstract

Big data refer to the massive datasets that are collected from a variety of data sources for business needs to reveal new insights for optimized decision-making. The solar and wind energy system is the modernization of electrical energy generation systems due to the pollution free nature and the continuous advancement of photo-voltaic and wind turbine system technologies. In the solar and wind energy surroundings, the application of big data analysis based decision-making and control are mainly in the following three aspects: data stream side management, storage side management and load side management. The objective of this research is to present a technological framework for the management of large volumes, variety, and velocity of solar system related information through big data tools such as Hadoop to support the assessment of solar and wind energy system. The framework includes a modeling of system, storage, management, monitoring and forecast based on large amounts of global and diffuse solar radiation and wind energy system. This chapter also includes market basket model, the concept of solar and wind depository and application of the Map Reduce algorithm.

Keywords: solar energy system, wind energy system, big data, Hadoop, Map Reduce

1. Introduction

Big data refer to the massive data set that are collected from a variety of data sources for implementing solar energy and wind energy system at a particular place and to reveal new insights for better decision-making. Based on different data analysis of the any study area, it is observed worldwide a lot of places are available where solar radiation and wind velocity available in abundance. The accessibility of non-conventional energy resources at any place is a significant feature to develop the solar energy and wind energy system for different purpose. Big data capable to generate values related to solar energy system and wind energy system for the storage and processing of very large quantities of information that cannot be analyzed with traditional computing techniques. Big data are categories into three part volume, velocity, and the variety and assess the pre-feasibility assessment with the help of these three features which is shown in **Figure 1**.

With phenomenal development in the field of electricity generation through renewable energy system, solar and wind power data sources have risen sharply. Exhaustive use of wind power big data can provide an effective way for safe operation of high quality power supply of a wind energy system. The effective goal

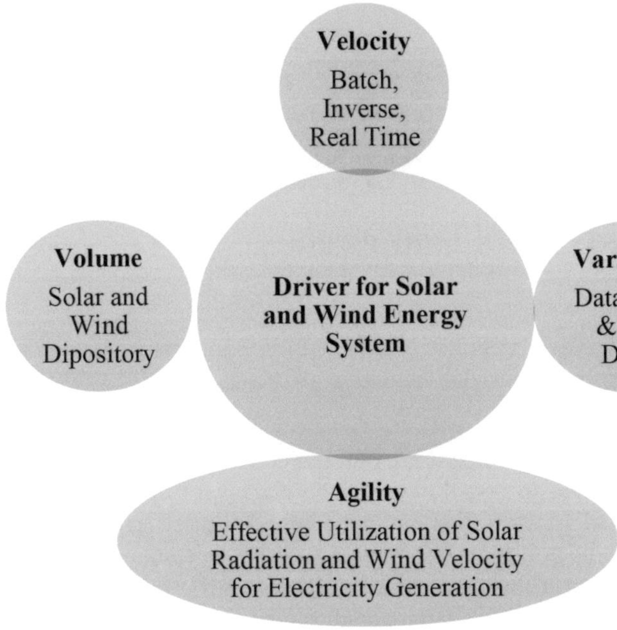

Figure 1.
Big data driver for solar and wind energy.

of big data research in the field of wind energy system is to "become aware of unawareness" and objectively there are well known facts and unknown facts which affect the working of wind energy system. Big data helps us in learning what we do not know and is done in two ways: from small to big and from big to small. It is necessary to develop a proper insight view of big data in the field of wind energy system and model out the wind system based on big data. Most big data applications in wind system consists two parts data corral and data dispel, in a data corral lot of information about the wind system project is collected and stored in wind cargo space, which is used to store the data and is utilized for further suitable application [1, 2].

Big data creates data from the processing of a large number of information which is related to digital information and such type of digital information cannot assess by normal computing processes. Wal-Mart handles near about 1.5 million consumer transaction in half an hour. Facebook handles 75 billion photos from its user base. Twitter develops 10 TB of data per day. By the big data analysis all of these processes are done in 1 week, which collect and process infinite number of data in a week, maybe without big data it was done in approximate 1 year. Hejazi et al. [3] described challenges and solution of electrical power system by big data analytics. In this paper different assessment method of power system is assessed by big data analytics. The paper also provides a holistic outline, classification and concise discussion on the technical approaches, research opportunity and application areas for energy big data analytics. Yao et al. [4] explained application of big data in the field of smart meter. In this article running path of smart meter is assessed by collection of lot of data, which is related to the operation and maintenance losses and their power information collection data. Shyam et al. [5] described assessment of smart grid through Apache spark based big data analytics. Apache Park is used to store the data related to the smart grid and respond according to the consumer requirements. Yang et al. [6] described different aspects and parameter of power to the gas energy system through big data analytics. In this paper big data assessment is also used for the operation and assessment of wind energy system.

Bersa et al. [7] predicted application of big data in power system. This paper also represented different technical parameter such as operational efficiency, losses and other parameter which is related to the power system is assessed by big data analytics. Percuku et al. [8] presented consumer load forecasting of electrical transmission system through big data analysis. The purpose of this article is to analyze a framework which is designed by Neo4j graph technology, which is a part of big data NoSQL data storage system. The Big data method is also used for prediction of load forecasting. Junaidi et al. [9] assessed electrical energy system through big data analytics. In this chapter, phasor measurement unit, modern days automated electric meter, and atmospheric measuring unit are assessed by big data analysis. Hangxun et al. [10] analyzed measuring power system quality by big data analytics. In this article a huge number of data is used to predict some power quality parameter such as sag and tension of power system, voltage, and current imbalance, which create lots of problems during the transmission and distribution of electrical energy. Huang et al. [11] analyzed different technical aspects of electrical energy system through big data. In this assessment author analyzed transient and steady state analysis of energy apparatus such as transformer, induction motor and synchronous condenser, etc. Wanxing et al. [12] described reactive power analysis of power system by big data topology. In this paper 8760 hours data of voltage and current by sending and receiving end is utilized for assessment of the reactive power of the electrical energy system. Zhan et al. [13] explained different prospects of smart grid or micro grid with the help of big data analytics. Guan et al. [14] described security and stability analysis of bulk power system. In this paper author assessed equal area criteria and other stability parameter with the help of huge number of data of load angle and consumer demand. Qing et al. [15] described impact of big data on electric power industry. The information in a specific sense has turned into another financial resource class. Step by step instructions to utilize big data to make more esteem will be another undertaking looked by all businesses, particularly the power business. In this paper, big data stage model of the power business and atomic power review and structure industry is planned.

This chapter is different from all above mentioned research in the application of big data in electrical power system. In this chapter, solar-wind energy system is analyzed through the big data analysis. Pre-feasibility analysis, modeling and financial analysis of solar-wind energy system is assessed by big data analysis. Market basket model and data retrieval is also used in this paper for assessment of solar-wind energy system.

2. Pre-feasibility assessment of solar-wind energy system by big data analysis

Electricity generation through solar and wind energy system is mainly depends on the amount of solar radiation and wind velocity at a particular site area. When the millions of solar radiation (kWh/m^2/day) and wind velocity (m/s) data are split into batches, sparse, interval and real time data, then this is the terms related to velocity of big data. When we consider a variety of big data on solar radiation and wind velocity data includes different types of data in the form of 3D data, audio, video, and unstructured text. When we store data in solar or wind depository, it is stored in the form of kilobytes, megabytes, gigabytes, terabytes, etc., and this represents the data volume of solar and wind energy system. For the feasibility assessment of solar and wind energy system in the study area data assessment of solar radiation, wind velocity, and hourly load consumption is categories in following two ways:

2.1 Modus operandi

Data that represents the real time status of solar data, wind data performance assessment, and loading of solar and wind energy equipment. This is the very basic information related to pre-feasibility assessment used by the system engineer to assess and manage the solar and wind power plant.

2.2 Non-modus operandi

A data file consists of data elements such as longitude, latitude and other data, and it also consists of a vendor specific property. Ancillary climatic data which is required in the database are in the form of air temperature, wind speed data, and water vapor content. Data of air temperature is used for calculating the PV module temperature, which is subsequently used to calculate PV power, after that wind cools the PV modules and modifies the temperature and hence the PV power. On the other hand, wind power is also depends on the specification of wind turbine, hub height, etc. **Figure 2** shows the number of applications of big data in solar and wind energy system.

The field of electricity generation through solar and wind energy source is the highly competitive world today and the downtime equates to real dollars lost and is deadly to company reputation. The NoSQL database environment is able to provide solar and wind system related data continue with operations without data loss and it is working like a solar and wind depository. Systems updates can be made dynamically without having to take the database offline. **Figure 3** shows the content and capacity of solar and wind depository. Solar and wind depository is divided into four parts for proper pre-feasibility assessment of solar and wind energy system and these are ERP, CRM, WEB and big data and data range consider from MB to PB. ERP module content data related to capital, replacement and operation and maintenance cost of photo-voltaic panel, generator, inverter, and battery system. CRM is the collection of nontechnical data which is related to the vendor information, client data, tender information, different types of solar and wind energy company data and data related to government policy, government subsidy, government renewable energy framework which is essential to develop a solar and wind energy system at study area. After the collection of all the three module information last one module is called big data module of solar and wind depository. It is a brain and heart of solar

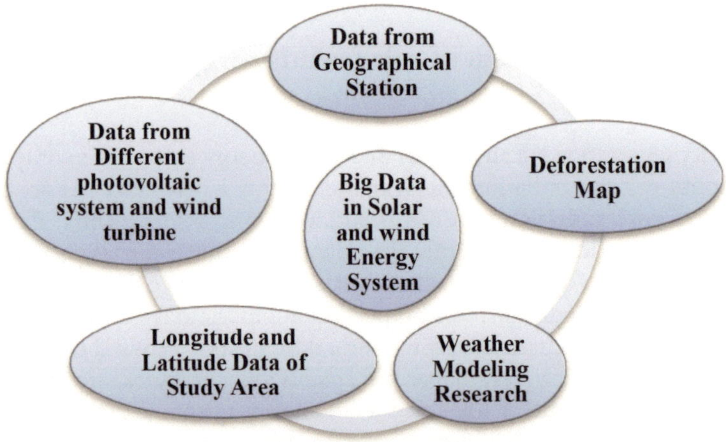

Figure 2.
Application of big data in solar and wind energy system.

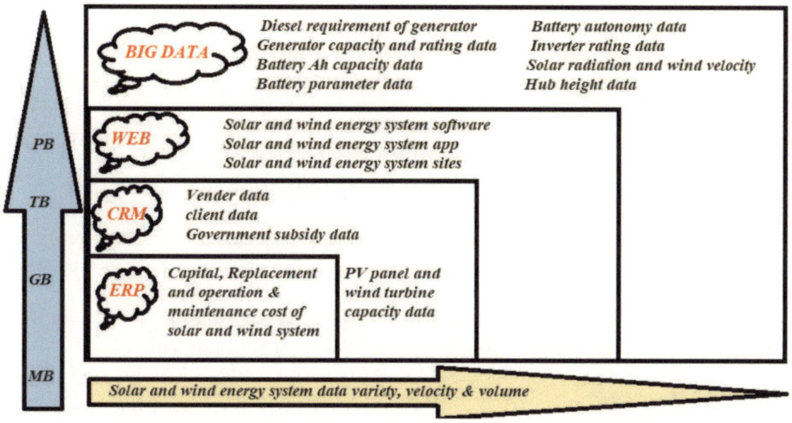

Figure 3.
Layout of solar and wind depository.

and wind depository because it accumulates all the data which is necessary to develop a solar and wind energy system in an efficient manner [16, 17].

The capability to read and write solar and wind system related information and in that position solar radiation and wind velocity input and electrical energy output physically occurs is termed as location transparency which is affected by longitude and latitude of the particular location. In a solar and wind depository document database key access pattern access data through a single key and navigates to another document through a related key.

3. Solar-wind energy system by Hadoop environment

Hadoop ecosystem is a framework of various types of complex and evolving tools and components. Some of these elements may be very different from each other in terms of their architecture. If we apply the concept of Hadoop ecosystem in the field of solar and wind energy system, then it is categorized into four types, data management, data access, data processing, and data storage. **Figure 4** shows Hadoop solar and wind energy ecosystem is also categorized into three types of pre-feasibility assessment in terms of longitude and latitude data assessment, solar radiation and wind velocity data assessment and load demand data assessment. In Hadoop system data management is done by Oozie, Chukwa, Flume, and Zookeeper, where Oozie is an open source Apache Hadoop service used to manage and process submitted tasks. Hadoop works by the divide and conquer approach. Once a problem is divided, it is approached and processed by using distributed and parallel processing technique across Hadoop cluster. Big data problems are approached with distributed applications and Zookeeper helps in coordinating all the elements of the distributed applications. Flume aids in transferring large amounts of data from distributed resources to a single centralized repository. It is robust and fault tolerant and efficiently collects, assembles and transfer data. Apache Chukwa is an open source information gathering framework for checking the enormous conveyed frameworks. Apache Chukwa is based over the Hadoop Distributed File System (HDFS) and Map/Reduce structure and acquires Hadoop versatility and heartiness. Apache Chukwa additionally incorporates a flexible and an incredible toolbox for showing; observing and dissecting results utilize the gathered information [18, 19].

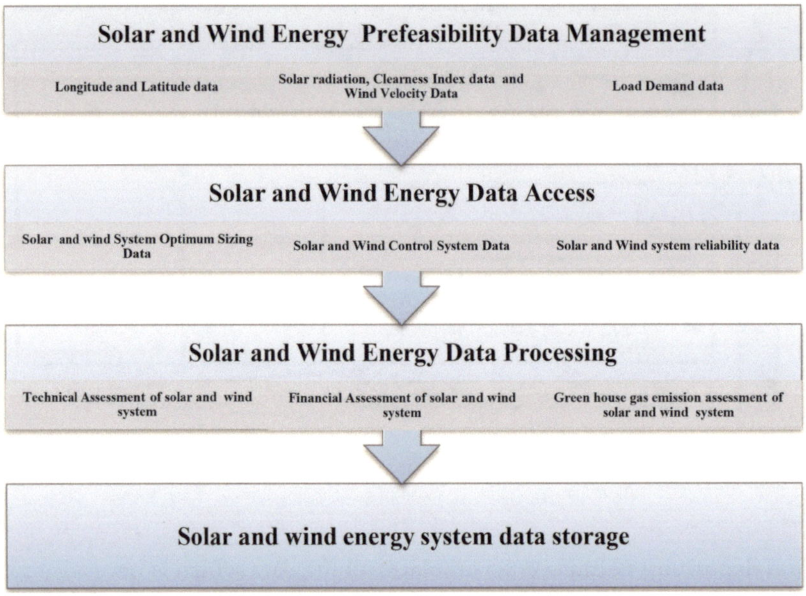

Figure 4.
Hadoop ecosystem for solar and wind energy system.

Data management of solar and wind energy system through the Hadoop system is always part of resource characterization and site assessment. In solar and wind energy system site examination is normally completed previously mentioned to utilization to set up the thorough physical encompassing for a specific sustainable power source venture, with the following objectives:

- To analyze the solar and wind energy generation life cycle of the given plan.
- To assess moderate global and diffuse solar radiation and high and low wind velocity conditions.
- To describe the clearness index and a hub height of the site area.

Objective and nature of resource assessment, solar and wind energy conversion characteristics, site condition, constraints, and physical boundary of assessment is the part of the solar and wind project description and which is the first step of resource allocation. If the project is in its initial stages and consists mainly of site screening, the resource assessment should be qualified as regional if the area of study is very large and incorporates many potential sites, whole country, or a large portion thereof.

4. Energy aware cluster node management of solar-wind energy system

Cluster node management is one of the parts of big data analysis and a cluster manager usually is a backend graphical user interface or command line software that runs on one or all cluster nodes. The cluster manager works together with a cluster management agent. A cluster is the process of making a group of abstract objects into classes of similar objects. The main advantage of clustering over classification is that, it is adaptable to changes and helps single out useful features that distinguish different groups.

Clustering method in solar and wind energy system:

- **Partitioning method**:

Suppose we are given a database of pre-feasibility assessment of solar and wind energy system which is represented by "n" database and the partition method construct 'k' partition of data. Then, a pre-feasibility assessment of the data is partially in the form of project description, estimation of current speed, result presentation, data analysis, and available and extractable energy data; so, "n" database is partitioned into k = 5 partitioned.

- **Hierarchical methods**:

In the hierarchical method of clustering, we identify step by step process of data gathering. In this method data collection and decomposition is done in two ways, first one is agglomerative and another one is divisive approach. In the agglomerative process merging the object and group that are close to one another and in the case of solar and wind energy system data are collected in the following manner:

1. Collection of pre-feasibility data of solar and wind energy system
2. Collection of data for modeling of solar and wind energy system
3. Collection of data for controlling of solar and wind energy system
4. Collection of data for reliability assessment of solar and wind energy system

In divisive approaches, we start with all of the objects in the same cluster and in the continuous iteration, a cluster is split into smaller clusters [20, 21]. According to the divisive approach agglomerative data is distributed and divide into following manner.

1. Collection of prefeasibility data of solar and wind energy system
    i. Location of the site
    ii. Geographical condition of the site
    iii. Data on solar radiation
    iv. Data on wind velocity
    v. Data of temperature
    vi. Data of rain fall
    vii. Data on consumer demand
2. Collection of data for modeling of solar and wind energy system
    i. Data of electricity required
    ii. Data of specification of solar panel and wind turbine
    iii. Data of specification of DC generator and AC generator
    iv. Data of specification of battery

3. Collection of data for controlling of solar and wind energy system

    i. Data of different types of errors

    ii. Data of different control strategies of solar and wind energy system

    iii. Data for stability analysis of solar and wind power plant

4. Collection of data for reliability assessment of solar and wind energy system

    i. Data of failure distribution model of solar and wind energy system

    ii. Data of time dependent failure model of a solar and wind energy system

    iii. Data of constant failure rate model

5. Basic big data measures for solar and wind data, text retrieval

In the text retrieval we need to check the accuracy of the data because it is very necessary part in the development of solar and wind power plant. Let the set of solar and wind energy system documents relevant to a query be denoted as (relevant solar and wind data) and the set of retrieved documents as (retrieved solar and wind data). The set of solar or wind documents that are relevant and retrieved can be denoted as.

$$(\text{Relevant solar or wind data}) \cap (\text{Retrieved solar or wind data})$$

This can be shown in **Figure 5** in the form of a Venn diagram as follows:

There are three fundamental measures for assessing the quality of solar and wind energy system retrieval:

- Precision
- Recall
- F-Score

Precision: precision is the percentage of solar and wind energy system retrieved documents that are in fact relevant to the consumer query. Precision of solar and wind power plant can be defined as:

$$\textit{Precision of renewable energy system data} = \frac{(\textit{Relevant solar} \vee \textit{wind data}) \cap (\textit{Retrieved solar} \vee \textit{wind data})}{(\textit{Retrieved solar} \vee \textit{wind data})}$$

Recall: recall is the percentage of solar and wind energy system documents that are relevant to the consumer query and were in fact retrieved. Recall is defined as:

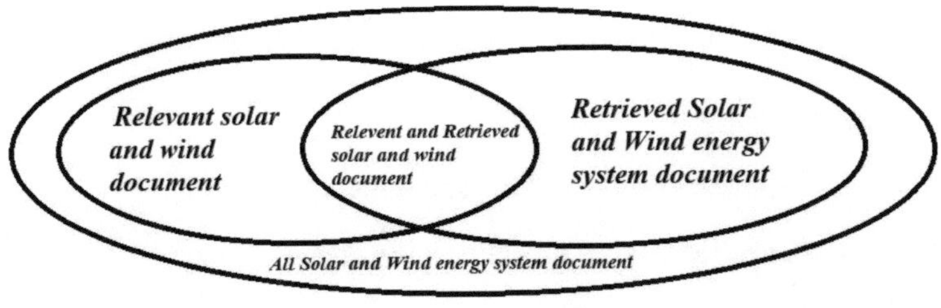

Figure 5.
Solar and wind text retrieval.

$$Recall\ of\ renewable\ energy\ system\ data = \frac{(Relevant\ solar \vee wind\ data) \cap (Retrieved\ solar \vee wind\ data)}{(Relevant\ solar \vee wind\ data)}$$

F-Score: F-score is the commonly used parameter of statistical analysis and the given observation retrieved system often needs to exchange information for precision or accuracy purpose. F score is defined as the harmonic mean of recall or precision as follows:

$$F\text{-}Score = \frac{Recall \times Precision}{0.5 \times (Recall + Precision)}$$

6. Application of Map Reduces in solar and wind energy system

The Map Reduce algorithm contains two important tasks shown in **Figure 6**, namely Map and reduce, which is used to provide essential framework for any task and also reduce the time interval of completion of any task.

- The essential modeling and framework is done by Mapper Class
- To reduce the number of steps for completion of any events is done by Reducer Class.

If we assess solar-wind energy system by Mapper Class, then first is carrying the input parameter of solar or wind energy system, then it analyze further provide framework according to the certain parameter and sorted according to the requirement. The output of Mapper class is used by the Reducer class as an input parameter of solar or wind energy system, which in revolves searches identical pairs and reduces them.

Sorting: sorting is the key step of Map Reduce algorithms, which is used to analyze the parameter according to the given constraints and manipulate the data according to the requirements. Map Reduce trappings sorting algorithm to without human intervention sort the final key-value pairs from the mapper by their keys.

- Sorting methods are the first step of the mapping class.
- In the second step tokenizing the parameter which collects from the first step.
- To accumulate transitional keys, the Mapper class is designed the framework by another comparator class.
- The position of transitional parameters for a given Reducer is routinely sorted by the Hadoop system to form parameters (K2, {V2, V2, ...}) before they are presented to the Reducer.

Searching: looking assumes a significant job in the Map Reduce calculation. It helps in the combiner stage (discretionary) and in the Reduce stage. Give us a

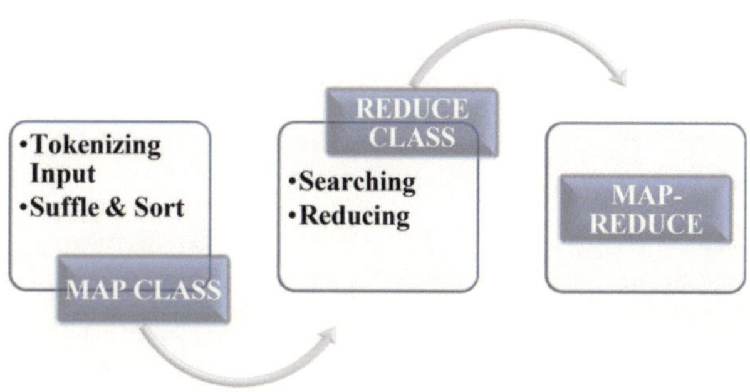

Figure 6.
Task of map-reduce algorithm.

chance to attempt to see how Searching functions with the assistance of a model. The following example shows how Map Reduce employs a searching algorithm to find out the details of the solar radiation and wind velocity and which country draws the highest solar radiation and wind velocity in a given atmospheric dataset.

- Let us assume we have solar radiation or wind velocity' data in four different files—W, X, Y and Z. Let us also assume that there are duplicate solar radiation or wind velocity' records in all four files because of importing the solar radiation or wind velocity data from all database tables repeatedly. See the following illustration and keep in mind value of solar radiation (kWh/m^2/day) and wind velocity (m/s) both are lies in the 0–1.

Country wise site area, average solar radiation or wind velocity	**Country wise site area, average solar radiation or wind velocity**	**Country wise site area, average solar radiation or wind velocity**	**Country wise site area, average solar radiation or wind velocity**
India, 5.2	China, 6	India, 5.2	India, 5.2
Japan, 5	Japan, 5	USA, 9	Japan, 5
Germany, 7	Germany, 7	Germany, 7	New Zealand, 8
Austria, 5	Austria, 5	Austria, 5	Austria, 5

The Map phase processes, each input file and provides the solar and wind energy system data in key-value pairs (<k, v>:<Site area, solar radiation or wind velocity>). See the following illustration.

Country wise site area, solar radiation or wind velocity	**Country wise site area, solar radiation or wind velocity**	**Country wise site area, solar radiation or wind velocity**	**Country wise site area, solar radiation or wind velocity**
<India, 5.2>	<China, 6>	<India, 5.2>	<India, 5.2>
<Japan, 5>	<Japan, 5>	<USA, 9>	<Japan, 5>
<Germany, 7>	<Germany, 7>	<Germany, 7>	<New Zealand, 8>
<Austria, 5>	<Austria, 5>	<Austria, 5>	<Austria, 5>

The combiner phase (searching technique) will accept the input from the Map phase as a key-value pair with site area and solar radiation or wind velocity. Using searching technique, the combiner will check all the country wise site area to find the highest solar radiation or wind velocity availability in each file. See the following snippet.

```
<k: country wise site area, v: solar current or wind velocity>
Max = the value of solar current or wind velocity of a first country. Treated as max solar current or
wind velocity
if(v(second country).solar_current or wind_velocity > Max)
{
  Max = v(solar_current or wind velocity);
}
else
{
  Continue checking;
}
```

The expected result is as follows:

<USA, 9> <New Zealand, 8> <Germany,7> <China, 5>

Reducer phase: form each file, you will find the highest solar radiation or wind velocity. To keep away from excess, check all the <k, v> matches and dispose of copy passages, assuming any. A similar calculation is utilized in the middle of the four <k, v> sets, which are originating from four info documents. The last yield ought to be as per the following:

<USA, 7>

Indexing: regularly indexing is utilized to point to a specific information and its location. It performs cluster ordering on the information records for a specific Mapper. The ordering strategy that is ordinarily utilized in Map-Reduce is known as reversed file. Web crawlers like Google and Bing utilize reversed ordering method. Give us a chance to attempt to see how Indexing functions with the assistance of a straightforward model. The following text is the input for inverted indexing. Here X [0], X[1], and X[2] are the file names and their solar radiation data are in double quotes [28].

X[0] = "5.2, 7.2, 6.3, 9.4"
X[1] = "9.4, 7.2, 8.1"
X[2] = "7.2, 6.3, 8.8, 4.5"

After applying the Indexing algorithm, we get the following output of solar radiation:

"7.2":{0,1,2}
"5.2":{0}
"6.3":{0,2}
"4.5":{2}

Here "4.5":{2} implies the term "a" appears in the X[2] file. Similarly, "7.2": {0, 1, 2} implies the term "is" appears in the files X[0], X[1], and X[2].

TF-IDF: TF-IDF is a substance getting ready estimation which is short for Term Frequency-Inverse Document Frequency. It is one of the fundamental web examination estimations. Here, the term 'repeat' suggests the events a term appears in a file.

Term frequency (TF): it gauges how much of the time a specific term in a record. It is determined by the occasions a parameter shows up in a report partitioned by the absolute number of parameters in that record.

TF(He) = (number of times term the '7.2' appears in a document)/(total number of terms in the document)

Inverse document frequency (IDF): it measures the noteworthiness of a term. It is controlled by the amount of reports in the substance database separated by the amount of chronicles where a specific term appears. While figuring TF, all of the terms are considered likewise huge. In this manner we have to know the regular terms while scaling up the uncommon ones, by finishing the accompanying:

IDF(He) = log_e(total number of documents/number of documents with term '7.2' in it)

The algorithm is explained below with the help of a small example.

In a particular site of India, in pre-feasibility assessment data containing 8760 data on solar radiation in a year, wherein solar radiation 7.6 $kWh/m^2/year$ appears 72 times in a year. Calculate the terms frequency and inverse document frequency of the solar data.

$$\text{Solution}: \textit{Terms Frequency}(7.6) = \frac{\textit{Number of} \times \textit{term the} \text{'}7.6\text{'} \textit{appears} \in \textit{a document}}{\textit{Total number of terms} \in \textit{the document}}$$

Terms frequency (7.6) = 72/8760 = 0.008

$$\textit{Inverse Document Frequency}(7.6) = \log_e \frac{8760}{72} = 2.08$$

7. Market basket model in solar and wind energy system

It shows that many relationships between two concepts "items" and "baskets" and each basket consist of a number of items. In the context of solar and wind energy system market basket model is considered in three ways, pre-feasibility assessment, modeling, and reliability assessment and it contain a lot of items which is related to the solar and wind energy system and shown in **Figures 7** and **8**. Basket of pre-feasibility assessment contains a lot of information because prior to installation and operation, the pre-feasibility study of solar and wind energy system should be done. In solar and wind energy projects an initial study undertaken to determine whether it is worthwhile to continue to the feasibility study stage. A precise feasibility study should provide a chronological background of the projects. In addition to climate condition of the application site, availability of solar and wind energy sources, the potential of solar and wind energy sources, load demand of application sites are included to find out the best location to develop a solar and wind renewable energy system. Generally feasibility precedes technical development and project implementation. It must therefore be conducted with a balanced approach to provide information upon which decisions can be based. Modeling basket contains different technique which is used to model the solar and wind energy system such as through HOMER, fuzzy logic and analytical technique. Modeling is the first step to design a system according to the different parameter and constraint. Modeling of solar and wind energy system is based on annual cost, battery autonomy function, sizing criteria and ecological statistical factor. Step by step optimization practice is used to find out the efficient result of the solar system model. Third and most important basket is reliability basket which is very important in the recent scenario because it evaluate the failure rate of individual component and overall solar energy system [22, 23].

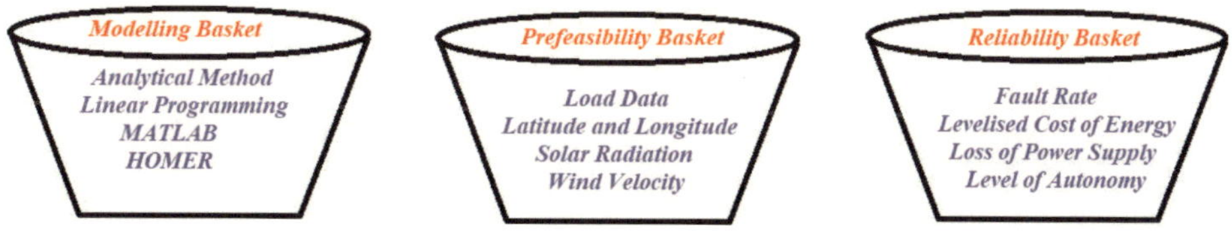

Figure 7.
Market basket of solar energy system.

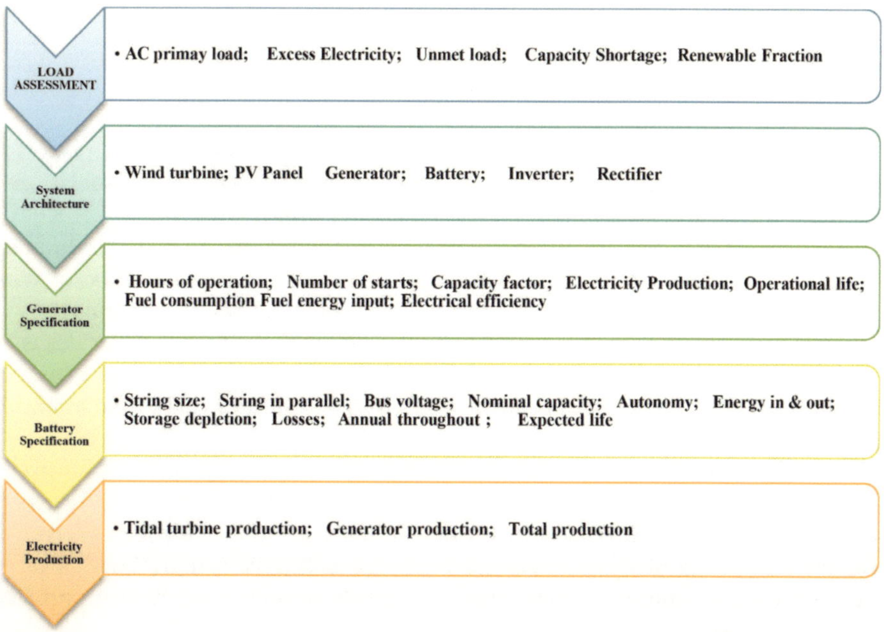

Figure 8.
Market basket model for optimum sizing and modeling of solar and wind energy system.

- **Market basket model based frequent item set mining for solar and wind energy system:**

Let K = {k_1............k_n} be a set of parameters. Let 'E' be the task relevant data, be a set of database parameter where each parameter T is a set of parameters such that $E \subseteq K$. Each parameter is associated with an identifier, called TKE. Solar and wind energy system parameter is referred to as a parameter set. We can call a parameter K a "frequent item set" only if its support count is sufficiently large. We prescribe a minimum support 's' and any K which has support greater than or equal to 's' is a frequent parameter set.

If we consider parameters of solar and wind energy system = {solar radiation (1), Sea surface temperature (2), Wind velocity (3), Temperature (4), Rain fall (5)}; Minimum support s = 3.

Transactions

1. T1 = {1,2,4}

2. T2 = {1,3,5}

3. T3 = {1,4}

4. T4 = {2,5}

5. T5 = {1,3,4}

6. T6 = {1,2,4,5}

7. T7 = {2,4,5}

8. T8 = {4,2}

Frequent parameters: {1}, {2}, {4}, {5}, {1, 4}, {2, 4}, {5, 2}

- **Associate rule mining in solar and wind energy system**:

The main purpose to discovering frequent parameters from a large dataset is to discover a set of "If-then" rules called Association rules. The form of association rules is K-j where K is a set of parameters of solar or wind power plant. Let K = {k_1, k_2.....k_n} be a set of n distinct attributes of solar and wind energy system which is also called literals of renewable power plants. Let 'E' be a database of solar or wind energy system, where each record of supply and demand side, T has a unique identifier, and contains set of parameters from the set K. An association rules is an implication of the form X → Y, where X, Y ⊆ K are item sets and X ∪ Y is a frequent item set. Thus the frequency of occurrence in X∪Y is at least equal to the minimum support s.

To test the reliability of solar or wind energy system, we try to define "confidence" of the rule which presents performance parameter of solar or wind power plant. Let X→Y be an association rule. The confidence of the rule which assess performance of solar or wind power plant defined as the fraction of the solar or wind power plant parameters that supports the rule among those that support the antecedent:

$$\textbf{Confidence}\ (\mathbf{X}{\to}Y) := \mathbf{P}\ (\mathbf{Y} \setminus \mathbf{X}) = \textbf{support}\ (\mathbf{X} \cup Y)/\textbf{support}\ (\mathbf{X})$$

The confidence of the rule indicates the degree of correlation between the certain parameter of solar or wind energy system. Such that value of solar radiation also depends on the value of wind velocity, so it is necessary to determine correlation between solar radiation and wind velocity.

Consider a small database with four parameters of combined solar and wind energy system K = {wind turbine specification (TT), Photo-voltaic Specification (B), Generator Specification (G), Consumer demand in Kw (CD)} and four transactions of these parameters shown in **Table 1**. **Table 2** shows all the parameter for K. Suppose that the minimum support and minimum confidence of an association rule are 40 and 60%, respectively which is shown in **Table 3**.

- **Framework for frequent parameter mining**:

The market baskets are also organized in memory. Generally, market basket data are stored in a file basket by basket. Generally, market basket data of solar and wind energy system are stored in a memory basket by basket. **Figure 9** shows Basket of Parameter Mining.

7.1 Monotonicity and Apriori algorithm property of solar and wind energy parameters

Given a database of transaction '**E**' over '**K**' and two sets X, Y⊆K, Then

Transaction ID	Parameters
T1	Tidal turbine, Photo-Voltaic, Generators
T2	Photo-Voltaic, Generators, Consumer Demand
T3	Photo-Voltaic
T4	Tidal turbine, Photo-Voltaic

Table 1.
Transaction of solar and wind energy system database.

Parameters	Support 's'	High/Low
Wind turbine	50%	High
Photo-Voltaic	100%	High
Generators	50%	High
Consumer Demand	25%	Low
Wind turbine, Photo-Voltaic	50%	High
Wind turbine, Generator	25%	Low
Wind turbine, Consumer Demand	0%	Low
Photo-Voltaic, Generator	50%	High
Photo-Voltaic, Consumer Demand	25%	Low
Generator, Consumer Demand	25%	Low
Wind turbine, Photo-Voltaic, Generators	25%	Low
Wind turbine, Photo-Voltaic, Consumer Demand	0%	Low
Wind turbine, Generators, Consumer Demand	0%	Low
Photo-Voltaic, Generators, Consumer Demand	**25%**	**Low**
Wind turbine, Photo-Voltaic, Generators, Consumer Demand	**0%**	**Low**

Table 2.
Support for parameters in table and large parameters with a support of 40%.

Rule	Confidence	Rule-hold
Wind turbine⇒ Photo-Voltaic	**100%**	**Yes**
Photo-Voltaic ⇒Wind turbine	**50%**	**No**
Photo-Voltaic⇒ Generators	**50%**	**Yes**
Generators ⇒Photo-Voltaic	**100%**	**No**

Table 3.
Confidence of some association rule where confidence interval = 60%.

$X, Y \subseteq K \Longrightarrow \text{support}(Y) \leq \text{support}(X)$.

In the Monotonicity property of support also allows us to compact the information about frequent solar and wind energy parameters and shown in **Table 4**. First, some definition is given below:

1. Solar or wind energy system parameters closed if none of its immediate parameter has the same count as the parameter.

2. Solar or wind energy system parameter is closed frequent if it is frequently and closed.

3. Solar or wind parameter is maximal frequent if it is frequent and none of its immediate superset is frequent.

Financial analysis of solar or wind energy system is consist three parameters = {Capital cost, Replacement cost, Operation and Maintenance cost} and the following baskets:

1. {Capital cost, Replacement cost}

2. {Capital cost, Replacement cost}

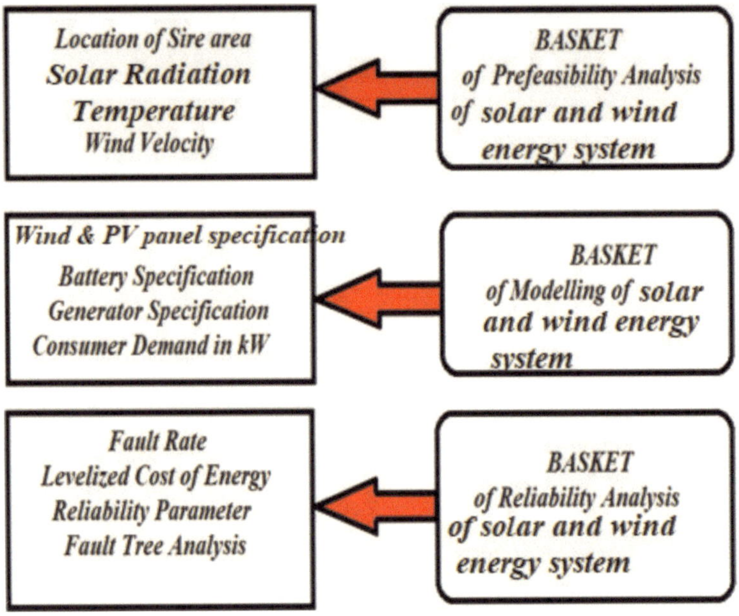

Figure 9.
Basket of parameter mining.

Parameter	Count	Frequent	Closed	Closed Frequency	Maximum Frequency
Capital cost	4	Yes	No	No	No
Replacement cost	5	Yes	Yes	Yes	No
Operation & Maintenance cost	3	Yes	No	No	No
Capital cost, Replacement cost	4	Yes	Yes	Yes	Yes
Capital cost, Operation & Maintenance cost	2	No	No	No	No
Replacement cost, Operation & Maintenance cost	3	Yes	Yes	Yes	Yes
Capital cost, Replacement cost, Operation & Maintenance cost	2	No	Yes	No	No

Table 4.
Indicating frequent, closed and maximal parameters.

3. {Replacement cost, Operation and Maintenance cost}

4. {Capital cost, Replacement cost, Operation and Maintenance cost}

5. {Capital cost, Replacement cost, Operation and Maintenance cost}

Assume the support threshold s = 3.

Monotonicity of solar and wind energy system is also done by Apriori algorithm and in this algorithm let's' be the minimum support required. Let 'n' be the number of items. In the first pass, we read the baskets and count and performance in main memory the occurrence of each parameter. In the second pass, we assess the basket again and count in main memory only those pairs where both parameters are frequent parameters. **Figure 10** shows solar-wind energy assessment through Apriori algorithm.

This algorithm is also used in solar and wind energy system for finding the most utilized parameter without counting all the necessary parameters can be extended to find larger frequent parameters without an exhaustive count of all data sets of tidal energy system. In the Apriori algorithm one step of each parameter of solar and wind energy system is taken for each set size is K [24, 25]. The pattern of moving from one size K to the next size K + 1 can be summarized as follows. For each size K, there are two sets of parameters:

1. DK is the set of parameter of size K, the solar or wind energy parameter that we must assess in order to determine whether they are in fact frequent.

2. MK is the arrangement of really visit parameters of size K.

The example of moving starting with one set, then onto the next and one size to the following is portrayed:

Assume we have assessed the reliability of solar or wind energy system and consider any parameters = {Fault rate (FR), Levelized cost of Energy (LCE), Loss of Power Supply Probability (LPSP), Level of Autonomy (LA), Minimum Time to Failure (MTF)}in the reliability basket.

1. {FR, LCE}

2. {FR, LCE, LPSP}

3. {FR, LCE, LA}

4. {LCE, LPSP, LA}

5. {FR, LCE, LPSP, LA}

6. {FR, LCE, LA, MTF}

Let the support threshold s = 3. The Apriori algorithm as follows:

1.a. Construct D1 = {{FR},{LCE},{LPSP},{LA},{MTF}}.

b. Assess the support of parameters of reliability assessment of solar or wind energy system in D1.

c. Remove infrequent parameter to get M1 = {{FR},{LCE},{LPSP},{LA}}.

2.a. Construct D2 = {{FR, LCE},{LCE, LPSP},{FR, LA},{LCE, LPSP},{LCE, LA}, {LPSP, LA}}.

b. Assess the support of parameters of reliability assessment of solar or wind energy system in D2.

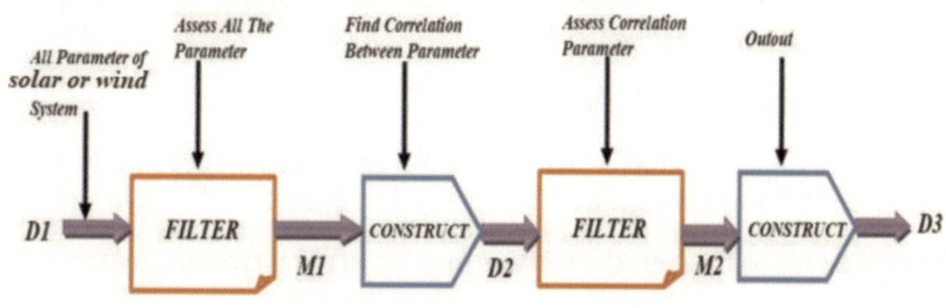

Figure 10.
Assessment through Apriori algorithm.

c. Remove infrequent parameter to get M2 = {{FR, LCE},{LCE, LPSP},{FR, LA},{LCE, LPSP}}.

3.a. Construct D3 = {{FR, LCE, LPSP},{FR, LCE LA},{LCE, LPSP, LA}}.

b. Assess the support of parameters of reliability assessment of solar or wind energy system in D3.

c. Remove infrequent parameter to get M3 = {{FR},{LCE},{LA}}.

4.a. Construct D4 = {Empty set}

Above assessment through market basket model shows that frequent item sets play an essential role in many data mining tasks of solar and wind energy system. The identification of different parameter, characteristics, and often together technical and financial assessment of solar and wind energy system is one of the basic tasks of the market basket model [26, 27].

8. Case study

In the modeling of solar-wind energy system, design a framework according to the certain parameter which is utilized for electricity generation and fulfill the consumer demand. A simulation of solar-wind energy system is an approximate imitation of the solar-wind energy system operation and developed the model with the certain boundary condition. The model of the solar-wind energy system is a well-defined description of the simulated parameter with key properties, such as technical, managerial, functional, and physical properties. Simulation through the data analysis is a key process in the recent scenario and we model out the system according to the certain parameter, where data follow the properties of big data such as volume, velocity, and variety.

In this modeling of solar-wind energy system we consider the peak load of 8KW at particular site area of India. **Table 5** shows data required for simulation of solar-wind energy system.

This information was tested each 1 hour for 365 days of a year. In an average day energy utilization is higher in the first part of the day from 5 A.M. to 9 A.M. furthermore, at night from 7 P.M. to 11:30 P.M. Numerical demonstrating is the initial phase in the structure of any solar energy sustainable power source framework and it gives an accurate perspective on any sustainable power source framework. If we consider randomly any site of particular countries and try to model out of solar-wind energy system so first necessary to define or assess certain parameter, which is essential to design the framework of solar-wind energy system. **Table 6** shows the necessary parameterization of a particular area.

According to the big data analysis following condition is satisfied for the above parameterization data:

- It should be necessary; we have all the above parameterization data in terms of terabyte or petabyte range.
- It should be necessary; all the data are well structured data.
- All the individual parameter data is structured in the form of hourly wise, day wise, month wise and year wise.
- It is necessary to find out the relationship in between two parameter.

Wind velocity of the study area is definitely affecting the clearness index and the amount of solar radiation. A relation between solar radiation and wind velocity and

Data Required for Simulation	Velocity	Volume	Variety
Year-wise solar radiation current data	Peta bytes	Stream	Structured
Year-wise wind velocity data	Peta bytes	Stream	Structured
Energy consumption data	Peta bytes	Stream	Structured
Optimum sizing of solar wind plant data	Peta bytes	Stream	Structured
Load demand data	Peta bytes	Stream	Structured

Table 5.
Data required for simulation.

clearness index and wind velocity is developed by regression analysis on the base of 8760 hours solar radiation, wind velocity, and clearness index data on the particular site. The relation is given by the equation:

$$SolarRadiation = 0.0039v^2 - 0.0029v + 5.9045\ (\textbf{According to the gathered data})$$

$$Cleanessindex = 0.0001v^2 - 0.0032v + 0.7643\ (\textbf{According to the gathered data})$$

The coastal vulnerability index (CVI) of the coastal area of given by:

$$\boldsymbol{CVI = 4G + 4S + 2C + 4T + 3W}\ (\textbf{According to the gathered data})$$

Parameter	Range
Latitude	23'10
Longitude	79'21
Solar Radiation	4 to 8kWh/m^2/day
Wind Velocity	5 to 25m/s
Hub Height	10m
Load Demand	05 to 2.5kW

Table 6.
Parameterization of site area.

VARIABLE STATISTICS	NO. OF OBSERVATION	Mode frequency	Categories	Frequency per category	Rel. frequency per category (%)	Lower bound on frequencies (95%)	Upper bound on frequencies (95%)	Proportion per category	Lower bound on proportions (95%)	Upper bound on proportions (95%)
RADIATION	4711	2827452	4	1631941.000	15.361	15.340	15.383	0.154	0.153	0.154
			5	2177457.000	20.496	20.472	20.520	0.205	0.205	0.205
			6	2448561.000	23.048	23.023	23.073	0.230	0.230	0.231
			7	2727462.000	25.673	25.647	25.700	0.257	0.256	0.257
			8	1638324.000	15.421	15.400	15.443	0.154	0.154	0.154
WIND VELOCITY	4711	1551197	8	857143.000	8.068	8.052	8.085	0.081	0.081	0.081
			9	1149660.000	10.822	10.803	10.840	0.108	0.108	0.108
			10	1441197.000	13.566	13.545	13.586	0.136	0.135	0.136
			11	1159172.000	10.911	10.892	10.930	0.109	0.109	0.109
			12	1438055.000	13.536	13.516	13.557	0.135	0.135	0.136
			13	1144192.000	10.770	10.751	10.789	0.108	0.108	0.108
			14	579460.000	5.454	5.441	5.468	0.055	0.054	0.055
			15	1419131.000	13.358	13.338	13.379	0.134	0.133	0.134
			16	1435735.000	13.514	13.494	13.535	0.135	0.135	0.135
Clearness Index	4711	1441377	0.25	881376.000	8.296	8.280	8.313	0.083	0.083	0.083
			0.3	443280.000	4.173	4.161	4.185	0.042	0.042	0.042
			0.35	440592.000	4.147	4.135	4.159	0.041	0.041	0.042
			0.4	884256.000	8.323	8.307	8.340	0.083	0.083	0.083
			0.45	443664.000	4.176	4.164	4.188	0.042	0.042	0.042
			0.55	443856.000	4.178	4.166	4.190	0.042	0.042	0.042
			0.6	885984.000	8.340	8.323	8.356	0.083	0.083	0.084
			0.71	444240.000	4.182	4.170	4.194	0.042	0.042	0.042
			0.72	444048.000	4.180	4.168	4.192	0.042	0.042	0.042
			0.8	442704.000	4.167	4.155	4.179	0.042	0.042	0.042
			0.9	442512.000	4.165	4.153	4.177	0.042	0.042	0.042
			1	1331377.000	12.532	12.512	12.552	0.125	0.125	0.126
			1.1	1326960.000	12.491	12.471	12.510	0.125	0.125	0.125
			1.2	885216.000	8.332	8.316	8.349	0.083	0.083	0.083
			1.3	441744.000	4.158	4.146	4.170	0.042	0.041	0.042
			1.4	441936.000	4.160	4.148	4.172	0.042	0.041	0.042

Table 7.
Variable statistics of renewable energy parameter.

where G is the vulnerability ranking of geomorphology, S is the site area, slope, C is the Shore Line Change, T is the wave velocity range and W is the significant hub height. According to the tool of regression analysis **Table 7** shows the variable statistics of tidal energy system by the regression analysis tool.

Data of above table shows that by the huge number of data, we can easily analysis prefeasibility assessment of solar-wind energy system and also predict frequency, lower and higher frequency of solar radiation, wind velocity and clearness index.

9. Conclusion

In case of big data volume, variety and velocity are the three main drivers that gave a new dimension to the way analytics of solar and wind energy system. Big data helps the solar and wind power system to create new growth opportunities and entirely new categories of the solar and wind system that can combine and analyze solar and wind industry data. Following are the concluding remarks of solar and wind energy assessment through the big data analytics:

- Atmospheric prediction is always depends on the lot of data which is maybe a day-wise, month-wise and year-wise and in the case of solar and wind energy system lot of data is required for pre-feasibility assessment. So that in pre-feasibility assessment big data analytics one of the best assessment methods.

- Market basket model based big data analytics provides plentiful information about the solar and wind energy system framework, buyers, and suppliers, customer preferences that can be captured and assessed.

- Hadoop framework is used for modeling of solar-wind energy system by the huge amount of data by dividing the data into a number of parts such as overall specification of energy system into the specification of photo-voltaic panel, wind turbine, generator, battery, etc.

- Map-reduce algorithm is used to find out optimum value of technical and financial parameters of solar and wind energy system by huge number of data collection.

Author details

Vikas Khare[1,2*] and Aaquil Bunglowala[3]

1 Department of Electrical Engineering, STME, NMIMS, Indore, India

2 Bureau of Energy Efficiency, India

3 School of Technology, Management and Engineering, NMIMS, Indore, India

*Address all correspondence to: vikaskharekhare@gmail.com

References

[1] Torrecilla JL, Romo J. Data learning from big data. Statistics & Probability Letters. 2018;**136**:15-19

[2] Eckroth J. A course on big data analysis. Journal of Parallel and Distributed Computing. 2018;**118**: 166-176

[3] Hejazi HA, Rad HM. Power system big data analytics: An assessment of paradigm shift barriers and prospectus. Energy Reports. 2018;**4**:91-100

[4] Yao HW, Wang XW, Wu LS. Prediction method for smart meter life based on big data. Procedia Engineering. 2018;**211**:1111-1114

[5] Shyam R, Bharati GHB, Kumar S. Apache spark based big data analytics plate form for smart grid. Procedia Technology. 2015;**21**:171-178

[6] Yang Z, Gao C, Zhao M. Utilizing big data to explore the running opportunity of power to gas in energy system. Energy Procedia. 2019;**158**:2341-2347

[7] Bersa RJ. Chapter 10: Future trends for big data application in power system. In: Big Data Application in Power System. Cambridge, USA: Elsevier; 2018. pp. 223-242

[8] Percuku A, Minkovska D, Stoyanova L. Big data and time series use in short term forecasting in power transmission system. Procedia Computer Science. 2018;**141**:167-174

[9] Junaidi N, Shaaban M. Big data application in electric energy system. In: IEEE International Conference on Computational Approaches in Smart System Design and Applications; 2018. pp. 1-4

[10] Hangxun T, Hongang W. Measuring system of power quality by big data analytics. In: IEEE International Conference on Cloud Computing and Big Data Analytics; Chengdu, China: 2018. pp. 248-252

[11] Huang J, Niu L, Zhan J. Technical aspects and case study of big data based conditioning monitoring of power apparatuses. In: IEEE PES Asia Pacific Power and Energy Engineering Conference; Hong Kong, China: 2014. pp. 1-4

[12] Wanxing S, Keyan L. The anomalous data identification study of reactive power optimization system based on big data. In: International Conference on Probabilistic Method Applied to Power System; Beijing, China: 2016. pp. 1-5

[13] Zhan J, Huang J. Study of the key technologies of electric power big data and its application prospects in smart grid. In: IEEE PES Asia Pacific Power and Energy Engineering Conference; Hong Kong, China: 2014. pp. 5-8

[14] Guan L, Zhang J. Enhancing security and resilence of bulk power systems by multisource big data learning. In: IEEE Power and Energy Society General Meeting; Chicago, USA; 2017. pp. 1-5

[15] Qing L, Boyu Z. Impact of big data on electric power industry. In: IEEE International Conference on Big Data Analysis; Beijing, China; 2017. pp. 460-463

[16] Cox DR. Big data: Some statistical issues. Statistics & Probability Letters. 2018;**136**:111-115

[17] Glushkova D, Jovanovic P. MapReduce performance model for Hadoop 2.x. Information Systems. 2017: 1-10. In Press

[18] Shankarmani R, Vijayalakshmi M. Big Data Analytics. 2nd ed. Onterio Canada: Wiley Publication; 2017

[19] Babcock B, Datar M. Maintaining variance and k-medians over data streams windows. In: Proc. ACM Symp. On Principles of Database System; 2004. pp. 234-243

[20] Aggarwal CC, Reddy C. Data Calculus: Algorithm and Applications. India: CRC Press; 2013

[21] Gracia H, Ullman J. Database System: The Complete Book. 2nd ed. Upper Saddle River, NJ: Prentice Hall; 2009

[22] Arghandeh R, Zhou Y. Big Data Application in Power Systems. 1st ed. Cambridge, USA: Elsevier; 2018

[23] Agrawal R, Imielinski T. Mining association between sets of items in massive datasets. In: Proc ACM SIGMOD International Conference on Management of Data; 1993. pp. 207-216

[24] Toivonen H. Sampling large data bases for association rules. In: International Conference on Very Large Databases; 1996. pp. 134-145

[25] Agneeswaran V. Big Data Analytics Beyond Hadoop. Pearson Education USA: 2014

[26] Dean J, Ghemawat S. MapReduce: Simplified data processing on large clusters. Communications of the ACM. 2008;**51**(1):107-113

[27] Philip C, Zhang CY. Data-intensive applications, challenges, techniques and technologies: A survey on big data. Information Sciences. 2014;**275**(Suppl. C): 314-347

[28] Available from: www.tutorials_point.com

Chapter 2

Technical Requirements for Connecting Solar Power Plants to Electricity Networks

Omar H. Abdalla and Azza A.A. Mostafa

Abstract

This chapter discusses basics of technical design specifications, criteria, technical terms and equipment parameters required to connect solar power plants to electricity networks. Depending on its capacity, a solar plant can be connected to LV, MV, or HV networks. Successful connection of a medium-scale solar plant should satisfy requirements of both the Solar Energy Grid Connection Code (SEGCC) and the appropriate code: the Electricity Distribution Code (EDC) or the Grid Code (GC) as the connection level apply. Connection of a large-scale solar plant to the transmission network should satisfy the requirements of both SEGCC and GC. For Small-Scale Photovoltaic (SSPV), the connection should satisfy both the SSPV Connection Code and the EDC. The objectives are to establish the obligations and responsibilities of each party; i.e. operators and all network users, thus leading to improved security, higher reliability and maintaining optimal operation. The technical specifications include permitted voltage and frequency variations in addition to power quality limits of harmonic distortion, phase unbalance, and flickers. Operational limits and capability requirements will be explained and discussed. Solar power grid connection codes of Egypt are explored first. Finally, brief comparisons of PV codes and related codes of UK, Germany, USA, and Egypt are presented.

Keywords: solar energy, PV power plants, grid connection codes, technical requirements and criteria, electricity networks, power quality

1. Introduction

The share of renewable resources for generating electric energy is increasing worldwide to cope with increasing demand. Current generation expansion plans of various countries expect increasing share of renewable energy resources in the electricity generation mix. By 2020, utilities set a target to reach a ratio of 20% renewable energy of the total energy required for electricity generation. Other utilities forecasted a higher share reaching about 50% by 2050. Wind energy and solar energy are the most promising resources and proven to be efficient in real applications with decreasing competitive costs of generated electric energy. The increasing share of renewable energies to be integrated to electric power systems has resulted in technical issues such as power quality requirements, capacity limits, safety measures, security, protection systems, synchronization process, lower system inertia, etc.

Electricity regulator authorities and electric utilities have issued necessary regulation rules for connecting sources of renewable energy to power networks at distribution and transmission levels according to the source capacity. A general overview of grid connection codes for integrating photovoltaic (PV) power plants to grids is presented in [1]. It presents a useful survey of grid codes, regulations, and technical requirements for connecting PV systems to low-voltage and medium-voltage networks, including issues of power quality and anti-islanding. An interesting guide dealing with PV interconnection requirements [2] has been developed and issued by the Interstate Renewable Energy Council, North Carolina Solar Center, USA. The guide covers all steps required for connecting a small-scale renewable energy system to the electricity network, including technical, contractual, rates, and metering issues. PV connection codes to medium-voltage power grid in Germany are discussed in [3]. A comparison of the processes of connecting PV systems in Germany and California is explored in [4]. Standards developed by the Institution of Engineering and Technology (IET) named "Code of Practice for Grid Connected Solar Photovoltaic Systems" are available in [5]. In South Africa, the National Energy Regulator has approved the "Grid Connection Code for Renewable Power Plants Connected to the Electricity Transmission System or the Distribution System" as detailed in [6]. Generally, utilities around the world either modify their grid codes to include technical requirements for integrating renewable energy resources to grids or issue separate but complementary codes for renewable resources.

This chapter describes the technical design specifications and criteria, technical terms, and equipment parameters for successful connection and operation of medium- and large-scale solar energy systems to the electricity networks in Egypt. The aim is to provide basic information and background on the technical design specification and criteria, in addition to technical terms and equipment parameters that are required to connect solar power plants to the electricity networks. Connection and successful operation of a solar power plant must satisfy the requirements of the Solar Energy Grid Connection Code (SEGCC) [7], and in the meantime the solar energy producer should comply with the requirements of the Electricity Distribution Code (EDC) [8]/Grid Code (GC) [9], according to the case of connection the MV distribution network/the HV transmission network.

The SEGCC specifies the special requirements for connecting both Medium-Scale Solar Plants (MSSPs) and Large-Scale Solar Plants (LSSPs) to the distribution networks or to the transmission network according to the capacity of the solar power plant. The capacity of MSSPs' range is from 500 kW to less than 20 MW. The LSSP range is greater than or equal to 20 MW. MSSPs may be connected either to the MV distribution networks or to the HV transmission networks. However, LSSPs are normally connected to the HV or extra-HV transmission networks. Successful integration of a MSSP shall comply with the technical requirements of both the SEGCC and the EDC, when connected to the distribution networks (or the GC when connected to the transmission network level). Similarly, the connection of a LSSP to the HV/EHV transmission networks shall satisfy the technical requirements of both the SEGCC and the GC. Technical requirements and terms stipulated in these codes should be clearly understandable in order to properly implement the rules and procedures of theses codes.

The EDC consists of the technical regulation rules and procedures to control technical and legal relationships between the licensed distribution system operator (DSO) and all users of the distribution network. The GC specifies the rules and procedures in order to control technical and legal relationships between the transmission system operator (TSO) and the users of the transmission network. The aim of the codes is to ascertain the obligations and responsibilities of each partner, i.e., TSO, DSO, and all users, namely, electricity producers, bulk-load customers, MV/

LV subscribers, etc. This will result in maintaining optimal power system operation, enhanced system security, and higher reliability.

The stipulated technical specifications of connecting MSSPs and LSSPs to the distribution networks or to the transmission network comprise the permitted limits of voltage and frequency variations in addition to power quality evaluation criteria such as limits of phase unbalance, limits of total and individual harmonic distortions, and limits of flicker severity. Operational limits and capability of solar power plants will be explained and discussed in this chapter.

It is important to mention here that the technical requirements for connecting small-scale photovoltaic (ssPV) systems to the low-voltage distribution networks are specified in the ssPV connection code [10]. Even though the ssPV code is considered to be all the complementary documents that involve compulsory requirements for a LV subscriber seeking installation of ssPV system, the subscriber shall also satisfy the technical requirements of the EDC. For more details, interested readers may refer to [11] for exploring technical background of connecting ssPV systems to LV distribution networks in Egypt.

The remainder of the chapter is structured as follows: Section 2 discusses briefly basic solar energy systems; Section 3 presents the codes of connecting solar power plants to electric grids in Egypt; Section 4 describes the technical requirements and criteria for connecting medium- and large-scale solar parks to the MV distribution networks or to the HV/EHV transmission networks; Section 5 briefly reviews terms and criteria of power quality referred to in the SEGCC; Section 6 presents comparisons of some rules of PV grid connection codes of three countries, namely, the UK, Germany, and Egypt; Section 7 summarizes the main conclusions and recommendations; and the Appendix at the end of the chapter lists the main IEC technical specification standards for solar park grid connection codes.

2. Solar energy: a brief introduction

Solar energy is the radiant light and heat from the Sun that is harnessed using solar heating, photovoltaics (PV), concentrated solar power (CSP), solar architecture, and artificial photosynthesis. Solar power is the conversion of the energy from sunlight into electricity, either directly using PV, indirectly using CSP, or a combination. The Sun is 1.3914 million km in diameter, and the radiated electromagnetic energy rate is 3.8 × 1020 MW. **Table 1** shows yearly renewable energy (RE) resources and human consumption. **Figure 1** shows the world annual solar insolation [12].

As shown in **Figure 1**, Egypt is one of the countries that possess the highest solar insolation. **Figure 2** shows the average direct solar radiation in $kWh/m^2/day$ in

Yearly RE resources and human use of energy (EJ)	
Solar energy	3,850,000
Wind energy	2250
Biomass energy (potential)	Circa 200
Primary energy use (in year 2016)	Circa 557
Electricity generation (in year 2016)	Circa 89
Exajoule (EJ) = 10^{18}, J = 278 TWh.	

Table 1.
Annual renewable energy resources and human use of energy.

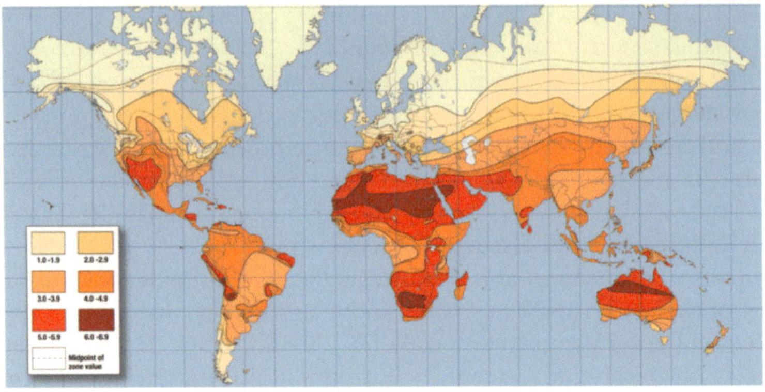

Figure 1.
Annual solar insolation worldwide [12].

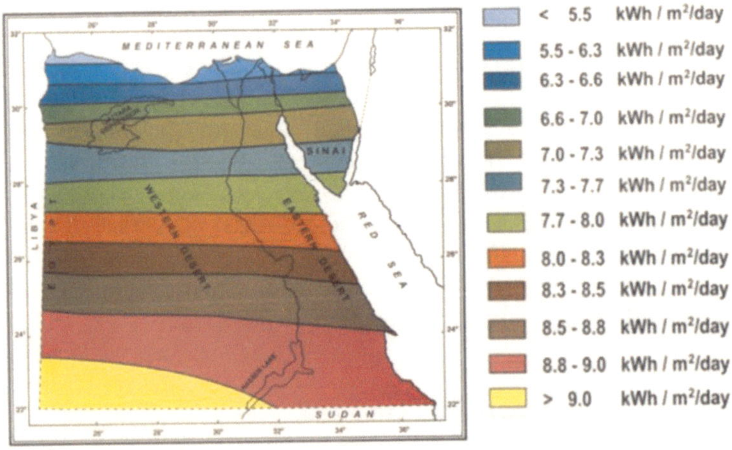

Figure 2.
Egypt solar atlas [13].

various regions in Egypt [11]. It can be noted that the southern regions have higher solar radiation than northern coastal regions. The region which has the highest solar radiation (>9.0 kWh/m^2/day) is shown in yellow in the figure.

Figure 3 shows the existing 1500 MW solar PV power plant located in Tengger Desert in China. It has been considered the largest PV power park in the world until now. Currently, Egypt is constructing a solar power plant of 1800/2000 MW in

Figure 3.
1500 MW Tengger Desert solar power plant in China.

Benban near Aswan [13]. It will comprise 40 PV stations of about 50 MW each. **Figure 4** shows an aerial view of part of the Benban PV solar power park [14]. Upon completion, Benban will be the worlds' largest PV power plant without energy storage.

Recent high concentration PV system is being developed by the IBM and the Air Light Energy Solutions using a parabolic dish to concentrate sunlight up to 2000 times onto new triple junction solar PV system. Each small (1 × 1 cm) chip can convert 50 W at 80% conversion efficiency, using liquid cooling process. **Figure 5** shows the concept of this new PV technology employing a tracking system to follow the sun.

Figure 6 shows the existing world's largest CSP plant (Ivanpah) located in California, in the Desert of Nevada in the USA. The installed capacity of this CSP plant is 392 MW [16]. The plant was commissioned in year 2014. Other larger CSP plants are currently under development in different countries. For example, Morocco's Ouarzazate solar power plant [17] will deliver about 580 MW of power once it is accomplished in year 2020. Also, Dubai authorities approved a CSP project to generate 1000 MW by 2020 and to be upgraded to 5000 MW by 2030.

Figure 7 shows the existing world's largest parabolic-trough solar energy generating systems located in Mojave Desert in California, USA. Its capacity is 354 MW

Figure 4.
Aerial view of under construction Benban PV power plant in Egypt [14].

Figure 5.
High-concentration PV system [15]. Image: www.airlightenergy.com/

Figure 6.
Ivanpah: the largest CSP plant in the world [16].

Figure 7.
Largest parabolic-trough concentrated solar system.

and includes 1600 acres. It was built in stages (1984–1990). The average capacity factor of this solar power plant is about 21%.

The concept of the solar updraft tower power plant (or solar chimney) [18] is shown in **Figure 8**. The solar chimney comprises four main parts, namely, the air collector, a tall tower, wind turbines, and an electric generator. The collector is suspended above the ground at a height of 2–20 m surrounding the tower. The solar radiation incident on the collector warms the air beneath the collector and makes it hotter than the outside air. The warmed air is drawn up through the tower, passing the wind turbine which is installed at the bottom of the tower base. The motion of air rotates the turbine and its associated electric generator.

Compared to PV systems, the solar chimney has the advantage of the possibility of operation 24 h a day even after sunset, thus overcoming the intermittency drawback of solar power. The available warm air beneath the collector can continuously operate the wind turbine and electric generator at night.

Figures 9 and **10** show the development of global solar energy generation from photovoltaic and concentrated solar power plants, respectively, up to year 2035 [19].

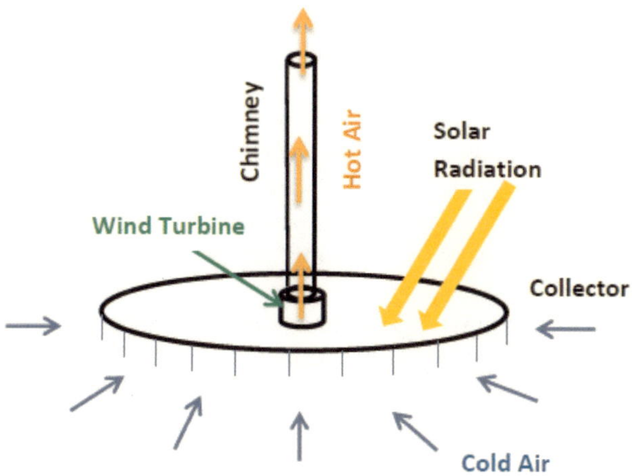

Figure 8.
Concept of solar chimney.

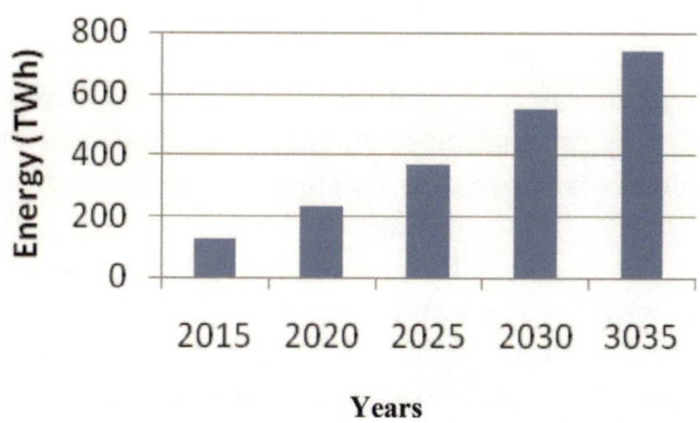

Figure 9.
Global energy generation from PV systems [19].

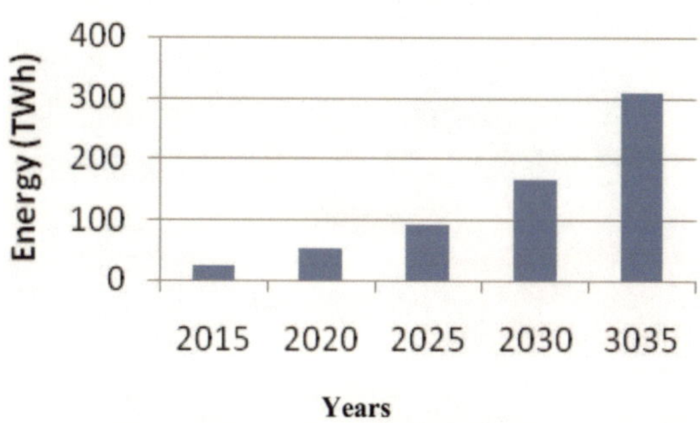

Figure 10.
Global energy generation from CSP plants [19].

3. Grid connection codes of solar power plants in Egypt

Two codes have been issued in Egypt for connecting solar power plants to electricity networks:

- The first one is ssPV code which stipulates the special requirements for the connecting small-scale photovoltaic systems (with rating < 500 kW) to low-voltage distribution networks [10].

- The second is the Solar Energy Grid Connection Code (SEGCC) which stipulates the technical requirements for connecting medium-scale (with capacity 500 kW to less than 20 MW) and large-scale (with capacity greater than or equal to 20 MW) solar power plants to the medium-voltage distribution networks or to the transmission grid.

The Grid Code (GC) in Egypt [9] defines the extra-high voltage (EHV) levels to be above 132 kV, the high voltage (HV) from 33 kV up to 132 kV, and medium voltage (MV) from 11 kV up to 22 kV. The solar plant grid connection codes are related to the following codes:

i. The Electricity Distribution Code (EDC) [8] which sets out the rules and procedures to regulate the relationship between the distribution utilities and users of the electricity distribution networks.

ii. The Egyptian Transmission System Code, commonly known as the "Grid Code" [9]. It sets out technical and legal relationships between the transmission system operator and the users of the transmission grid. The users are electricity production companies, distribution system companies, and bulk customers who are directly supplied from the transmission grid, etc.

In addition to the above codes, there is the "Wind Farm Grid Connection Code" [20] which concerns with the rules and procedures for connecting wind energy conversion systems to the transmission grid. The above five codes are shown in **Figure 11**. For instance, the wind grid farm connection code and the Grid Code are two complementary codes that should be fulfilled for connecting a wind farm to the transmission system.

The solar energy code and the Grid Code are two complementary technical documents that should be satisfied for connecting a solar power plant to the grid. The aim of the solar energy grid connection code is to stipulate the technical

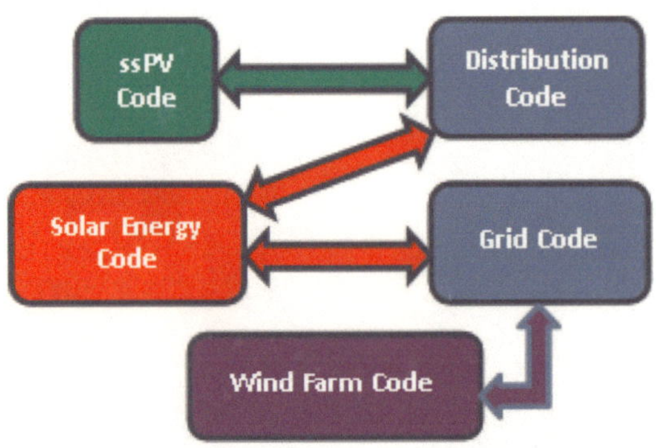

Figure 11.
Association of various codes in Egypt.

requirements for connecting solar energy resources either new or modified to the grid, so that security and quality of the grid are guaranteed.

The solar energy grid connection code specifies the special requirements for connecting solar energy plants to the MV distribution networks or HV/EHV transmission network. The technical requirements include permitted limits of voltage and frequency variations in addition to power quality limits such as of phase unbalance limits, harmonic distortion limits, and flicker severity limits. The code specifies also the operational limits of solar power plants to be integrated into the grid, plant capability requirements, active and reactive power control systems, safety measures, protection settings, synchronization, etc. The solar energy connection code shall apply to all medium-scale and large-scale solar power plants (either PV parks or solar thermal power plants) to be connected to the transmission grid. For connecting small-scale PV systems with capacity <500 kW to the LV distribution networks, we refer the reader to the small-scale PV (ssPV) code [10].

4. Solar energy grid connection requirements

4.1 Point of common coupling

The "point of common coupling (PCC)" is a point at which solar power plant is connected to the grid. It is sometimes called the "grid connection point (GCP)." The PCC is usually the connection point at the high-voltage terminals of the generator step-up transformer; it is generally located at the grid side of the isolating switch between the solar power plant and the grid. Normally, the solar energy grid connection code specifies the following technical requirements at the PCC.

4.2 Range of voltage

The grid-connected solar power plant shall be able to deliver its actual active power when the voltage at the point of common coupling remains within the ranges shown in **Table 2**. If required by the transmission system operator, the solar plant shall be also capable of automatically disconnecting from the grid at specified voltages.

4.3 Frequency range

In the case of a deviation of the grid frequency from its permissible value, the solar power plant shall perform as follows:

a. If the frequency is <50 Hz, the solar plant shall continue injecting active power until the frequency reduces below 47.5 Hz.

b. For over-frequency between 50 and 50.2 Hz, the solar power plant shall maintain the 100% of active power.

Range of voltage (pu)	Time of operation
0.85–1.10	Unlimited
1.10–1.15	30 min

Table 2.
Range of voltage at the PCC.

c. If the frequency is >50.2 Hz, the solar power plant shall inject active power up to 51.5 Hz.

4.4 Starting up solar power plants

The solar power plant shall only be connected to the power grid if the frequency and the voltage at the PCC are within the limits given in **Table 3** or as otherwise stated in the Connection Agreement (CA) between the transmission system operator and the owner of solar power plant.

During the start-up of a solar power plant, the active power increasing rate shall not exceed 10% (of the rated active power of the plant) per minute.

4.5 Power quality requirements

The solar plants connected to the power grid shall endeavor to maintain the quality of the voltage waveform at the PCC. The solar power plants shall comply with the requirements specified in Section 5.3 of the Performance Code of the Grid Code and/or the related part in the Electricity Distribution Code.

4.6 Harmonic distortion

The maximum harmonic distortion levels at the PCC which are attributable to the solar power plant shall obey the stipulations in the IEEE Standard 519-1992 as specified in Section 5.3.7 of Performance Code and/or the applicable section in the Electricity Distribution Code.

It is well known that a linear load, such as incandescent lamps or heaters, draws electric current from the source proportional to the applied voltage, while a nonlinear load such as an adjustable-speed drive draws currents apart from the voltage wave. The current of the nonlinear load comprises odd harmonics (third, fifth, seventh, etc.). The distortion effect of the third harmonic component is shown in **Figure 12**. Components of harmonic currents will interact with source currents, thus causing voltage harmonics. The voltage harmonic components are superimposed on the fundamental voltage component leading to a distorted voltage waveform. It may be mathematically described by the Fourier form Eq. (1):

$$f(t) = \alpha_o + \sum_{n=1}^{\infty} \alpha_n \cos(n\omega_o t) + \sum_{n=1}^{\infty} b_n \sin(n\omega_o t) \tag{1}$$

where

$$\alpha_0 = \frac{1}{T}\int_0^T f(t)dt + DC\ component \tag{2}$$

$$\alpha_n = \frac{2}{T}\int_0^T f(t)\cos(n\omega_o t)\,dt \tag{3}$$

Frequency	48.0 Hz ≤ f ≤ 51.0 Hz
Voltage	0.90 u ≤ U ≤ 1.10 pu

Table 3.
Limits of voltage and frequency during the start-up of a solar plant.

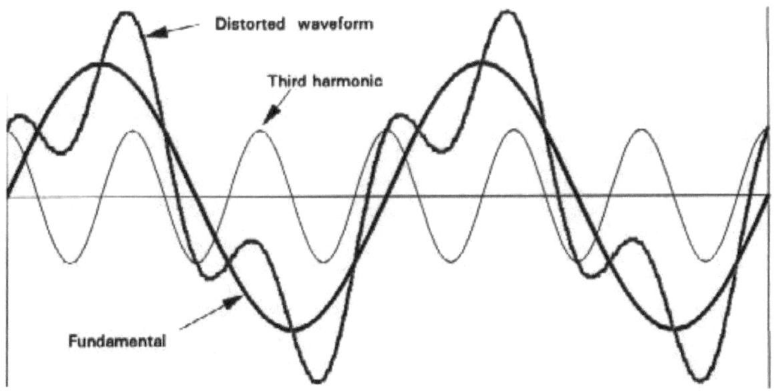

Figure 12.
Effect of the third harmonic.

$$b_n = \frac{2}{T}\int_0^T f(t)\sin(n\omega_o)t\,dt \tag{4}$$

The total harmonic distortion in voltage (THDv) and current (THDi) are defined as follows:

$$THDv = \frac{\sqrt{V_2^2 + V_3^2 + V_4^2 + V_5^2 + \dots.}}{V_1} \tag{5}$$

$$THDi = \frac{\sqrt{i_2^2 + i_3^2 + i_4^2 + i_5^2 + \dots}}{I_1} \tag{6}$$

The flow of harmonic currents in electrical equipment can cause problems such as heating of equipment, overloading neutral line, wrong tripping of circuit breakers, increasing skin effect, etc. Hence, electricity codes specify appropriate limitations on the total and individual harmonics in the grids. The solar energy grid connection code defines the limits of the individual and total harmonic distortion of voltage and current waveforms at the PCC as listed in **Tables 4–7** in accordance with the IEEE Standard 519-1992. The updated version of this standard (IEEE Standard 519-2014) has introduced new two rows as given in **Tables 4** and 7. We recommend using the updated version of the standard.

It should be noted that the harmonic distortion level may exceed the levels listed in the above tables for a period no longer than 30 s provided that such increases in harmonic distortion level do not compromise service to the users or cause damage to any equipment in the grid as determined by the TSO.

Level of voltage	Harmonic voltage distortion level (%)	
	Odd harmonic limits	Total harmonic limits
V ≤ 1 kV	5.0	8.0
1 kV < V ≤ 69 kV	3.0	5.0
69 kV < V ≤ 161 kV	1.5	2.5
V > 161 kV	1.0	1.5
The first row for (V ≤ 1 kV) has been introduced in the IEEE Standard 519-2014.		

Table 4.
Limits of harmonic voltage distortion.

Short circuit ratio	Maximum integer harmonic current distortion as percentage of I_L					
	Odd harmonic distortion**					TDD
I_{SC}/I_L	<11	≥11 to <17	≥17 to <23	≥23 to <35	≥35	
<20*	4.0	2.0	1.5	0.6	0.3	5
20 < 50	7.0	3.0	2.5	1.0	0.5	8
50 < 100	10.0	4.5	4.0	1.5	0.7	12
100 < 1000	12.0	5.5	5.0	2.0	1.0	15
>1000	15.0	7.0	6.0	2.5	1.4	20

where I_{SC} = the maximum short-circuit current at the PCC; I_L = the maximum demand load current (fundamental frequency component) at the PCC.
**All power generation equipment is limited to these values of current distortion, regardless of actual I_{SC}/I_L.*
***The limits of even harmonics are 25% of the corresponding limits of odd harmonics listed in the table.*

Table 5.
Harmonic current distortion for transmission voltage level 69 kV and below.

Short circuit ratio	Maximum integer harmonic current distortion as percentage of I_L					
	Odd harmonic distortion**					TDD
I_{SC}/I_L	<11	≥11 to <17	≥17 to <23	≥23 to <35	≥35	
<20*	2.0	1.0	0.75	0.3	0.15	2.5
20 < 50	3.5	1.75	1.25	0.5	0.25	4
50 < 100	5.0	2.25	2.0	0.75	0.35	6
100 < 1000	6.0	2.75	2.5	1.0	0.5	7.5
>1000	7.5	3.5	3.0	1.25	0.7	10

where, I_{SC} = the maximum short-circuit current at the PCC; I_L = the maximum demand load current (fundamental frequency component) at the PCC.
**All power generation equipment is limited to these values of current distortion, regardless of actual I_{SC}/I_L.*
***The limits of even harmonics are 25% of the corresponding limits of odd harmonics listed in the table.*

Table 6.
Harmonic current distortion for transmission voltage level above 69 kV up to 161 kV.

Short circuit ratio	Maximum integer harmonic current distortion as percentage of I_L					
	Odd harmonic distortion**					TDD
I_{SC}/I_L	<11	≥11 to <17	≥17 to <23	≥23 to <35	≥35	
<25*	1.0	0.5	0.38	0.15	0.1	1.5
<50	2.0	1.0	0.75	0.3	0.15	2.5
≥50	3.0	1.5	1.15	0.45	0.22	3.75
The first row for (<25*) has been added in IEEE Standard 519-2014						

where, I_{SC} = the maximum short-circuit current at the PCC; I_L = the maximum demand load current (fundamental frequency component) at the PCC.
**All power generation equipment is limited to these values of current distortion, regardless of actual I_{SC}/I_L.*
***The limits of even harmonics are 25% of the corresponding limits of odd harmonics listed in the table.*

Table 7.
Harmonic current distortion for transmission voltage level above 161 kV.

It should be also noted that the updated version IEEE Standard 519-2014 specifies the width of the window for measuring the harmonics to be 10 cycles in the 50 Hz systems, i.e., 200 ms window, as follows:

- For *very-short time* harmonic measurements, use the following equation:

$$F_{n,\,vs} = \sqrt{\frac{1}{15}\sum_{i=1}^{15} F_{n,\,i}^{2}} \tag{7}$$

- For *short time* harmonic measurements, use the following equation:

$$F_{n,\,sh} = \sqrt{\frac{1}{200}\sum_{i=1}^{200} F_{(n,vs),\,i}^{2}} \tag{8}$$

The system owner/operator should limit the line-to-neutral voltage harmonics at the PCC as follows:

- The values of the daily 99th percentile very-short time (which is 3 s in the 50 Hz systems) should be <1.5 times the values given in the tables.

- The values of the weekly 95th percentile short time (10 min) should be less than the values given in the tables.

For the current harmonic distortion **Tables 5–7**, the following points are applicable:

- The daily 99th percentile very-short time harmonic currents should be <2 times the values listed in the tables.

- The weekly 99th percentile short time harmonic currents should be <1.5 times the values given in the tables.

- The weekly 95th percentile short time harmonic currents should be less than the values given in the tables.

4.7 Limits of flicker severity

Table 8 shows the limits of the flicker severity produced by a solar energy power plant at the PCC as per recommendations of the IEC 61000-3-7.

Voltage flicker at the PCC is produced by voltage variations caused by a load such as an arc furnace when spectral characteristics of the voltage variations is in the range of a fraction of a cycle per second to about one third of the system frequency. It is a characteristic where a high-frequency (ω_o) sinusoid is modulated by a low-frequency sinusoid (ω_f).

In mathematical form

$$v(t) = \lfloor 1 + V_f \cos\left(\omega_f t\right) \rfloor \, V_m \cos\left(\omega_0 t\right) \tag{9}$$

Short-term (10 min)	$P_{st} \leq 0.35$
Long-term (2 h)	$P_{lt} \leq 0.25$

Table 8.
Levels of flicker severity at the PCC.

Intensity of flicker is given by

$$F = \frac{V_f}{V_m} = \frac{S_{scf}}{S_{sc}} \tag{10}$$

where S_{scf} is the short-circuit power (in MVA) at the electrode tip; S_{SC} is the short-circuit power (in MVA) at the PCC.

A flicker meter has been developed by the IEC to measure flicker severity in terms of fluctuating voltage magnitude and its corresponding frequency of fluctuations. The meter employs a software technique to convert measured voltage fluctuations to the following statistical quantities:

- Short-term flicker severity (P_{ST})
- Long-term flicker severity (P_{LT})

The flicker meter takes measurements automatically at 10-min intervals. The P_{ST} is calculated every 10 min. The flicker severity indicator P_{ST} which has a value of 1 is the level of visual flicker severity at which 50% of people would perceive flicker in a 60 W incandescent lamb. The long-term flicker severity P_{LT} is a combination of 12 P_{ST} measurement values of 10 min each.

4.8 Limits of voltage unbalance

The voltage unbalance in the three-phase system is defined as the difference between the highest and lowest line voltage divided by the average line voltage of the system. Solar power plants shall be able to withstand voltage unbalance not exceeding 2% for at least 30 s as stipulated in part 5.3.5 of Section 5 (Performance Code) of the Grid Code and/or the relevant section in the Distribution Code.

A three-phase system is balanced if the three-phase voltages have the same amplitude and are phase-shifted by 120° with respect to each other. Otherwise, the three-phase system is unbalanced. **Figure 13** shows the voltage waveforms of an unbalanced three-phase system.

The mathematical relationships between the symmetrical components of system voltages ($V_O - V_1 - V_2$) and the phase components ($V_A - A_B - V_C$) are given in Eqs. (11) and (12):

$$\begin{bmatrix} V_O \\ V1 \\ V2 \end{bmatrix} = \frac{1}{3} \begin{bmatrix} 1 & 1 & 1 \\ 1 & a & a^2 \\ 1 & a^2 & a \end{bmatrix} \begin{bmatrix} V_A \\ V_B \\ V_C \end{bmatrix} \tag{11}$$

$$a = e^{j120} \tag{12}$$

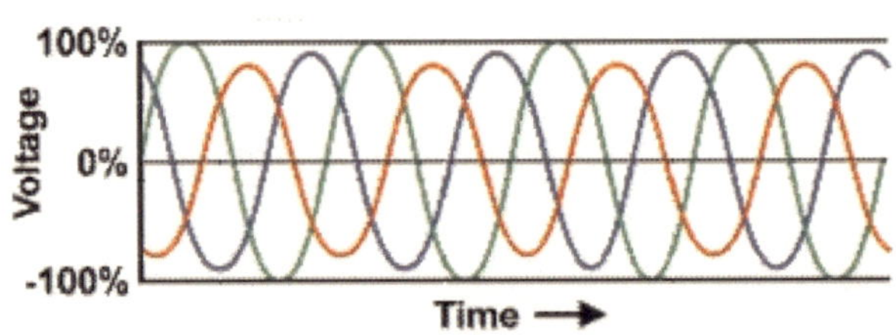

Figure 13.
Voltage waveforms of unbalanced three-phase system.

V_0 = zero-sequence component; V_1 = positive-sequence component; V_2 = negative-sequence component.

According to the EN-50160 and IEC-61000-3-x Standards, the voltage unbalance (V_{2U}) is defined as

$$V_{2U}\% = \frac{V_2}{V_1} \times 100 \tag{13}$$

The above standards define the following limits of voltage unbalance:

$$V_{2U} < 1\% \text{ for HV} \tag{14}$$

$$V_{2U} < 2\% \text{ for MV\&LV} \tag{15}$$

The voltage unbalance is measured as 10-min average value with an instantaneous maximum of 4%. Voltage unbalance may also be defined [21]:

IEEE definition of voltage unbalance

$$\%P_{vu} = \frac{\text{Maximum deviation from average } V_{ph}}{\text{Average } V_{ph}} \tag{16}$$

In Eq. (16) only magnitudes are considered.

NEMA defines the same formula but considers line voltages.

Approximate formula

$$\%VU = \frac{82 \times \sqrt{V^2abe + V^2bce + V^2cae}}{Average\, V_{line}} \times 100 \tag{17}$$

Subscript e means deviation from average. The causes of unbalance include generators; transformers; unbalanced impedances of long, non-transposed low-voltage lines; unbalanced load currents; single-phase loads on three-phase systems; etc. Unbalance can adversely affect motors and transformers by increasing heat and reducing their efficiencies.

4.9 Limits of voltage fluctuations

Voltage fluctuations, at the PCC of a solar power plant, can occur due to switching operations inside the solar plant elements such as transformers, capacitor banks, connection circuit, etc., resulting from inrush currents. These voltage fluctuations shall be up to 3% of nominal voltage provided that the fluctuations do not compose any risk to the grid or other connected users in the view of the TSO.

4.10 Control of active power

Figure 14 shows the ranges of voltage, frequency, and time periods within which the solar power plant shall continue delivering actual active power to the grid at the PCC. For grid frequencies in the range from 50.2 to 51.5 Hz, the solar power plant should reduce its active output power consistent with Eq. (18) and **Figure 15** providing that the voltage is within the range 0.9–1.1 pu:

$$\Delta P = 0.4 \times PM \times \Delta F \text{ perHZ} \tag{18}$$

where PM is the actual output power before the frequency of the grid exceeds 50.2 Hz; ΔF is the actual frequency minus 50.2 Hz.

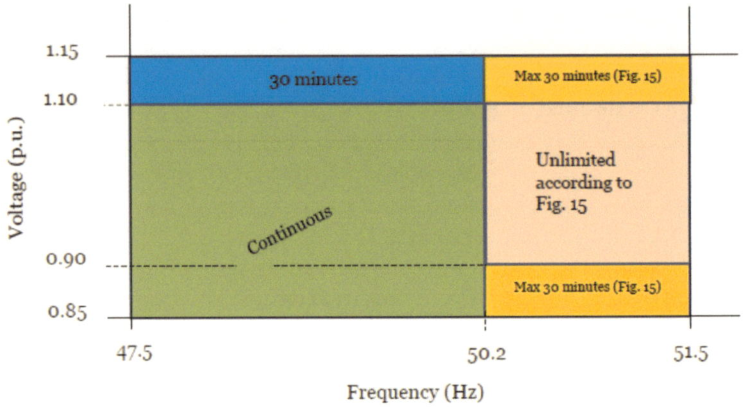

Figure 14.
Voltage, frequency, and time ranges of solar plant operation.

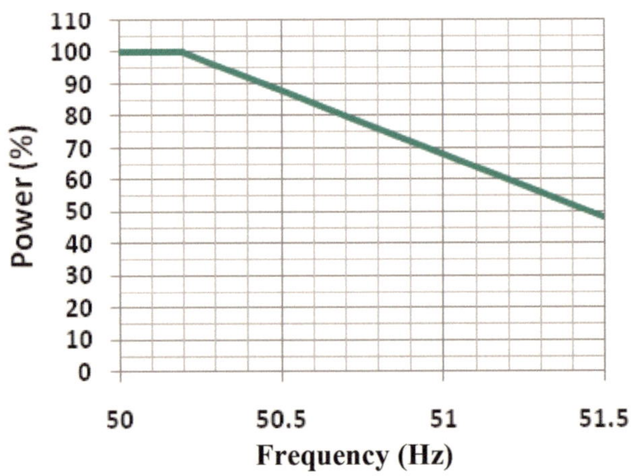

Figure 15.
Reduction in active power due to over-frequency.

Also, in this frequency range (i.e., 50.2–51.5 Hz) and the voltage ranges (0.85–0.9 pu) or (1.1–1.15 pu), the operation with reduced active power shall be limited to 30 min. The increasing or decreasing ramp of power will be performed in steps of a 10% (each) of the maximum power.

4.11 Control of reactive power

The solar power plant must be able to control reactive power at the PCC in a range of 0.95 lagging power factor to 0.95 leading power at the maximum active power of the plant and in consistent with **Figure 16** for the MSSPs and **Figure 17** for the LSSPs. The solar power plant must be able to perform reactive power control as follows:

- Set-point control of reactive power (Q)
- Set-point control of power factor
- Fixed power factor

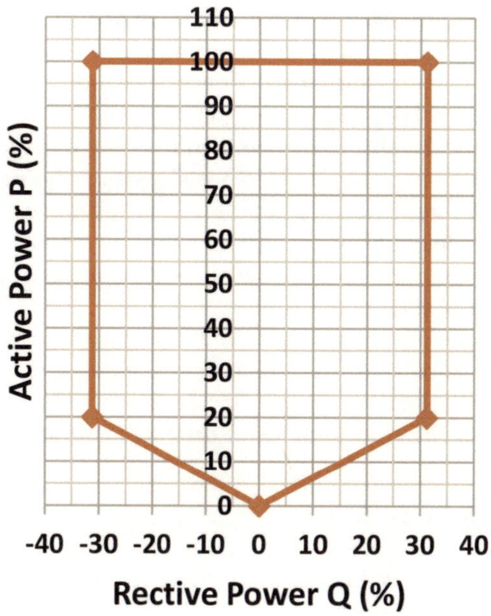

Figure 16.
P-Q capability chart for MSSPs.

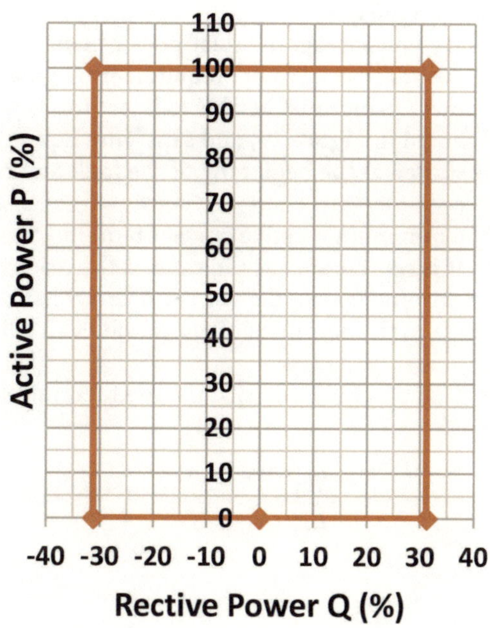

Figure 17.
P-Q capability chart for LSSPs.

- Characteristic: power factor as a function of active power output of the solar power plant, i.e., $cos\ \varphi\ (P)$

- Characteristic: reactive power as a function of voltage, i.e., Q (V)

The solar power plant must possess an input signal for a set-point value at the PCC in order to control the reactive power or power factor of the plant. It is able to receive the set point within reactive power accuracy of 1 kVAr. The set-point signal will be provided by the TSO through verbal communication or SCADA, whichever is available. The solar power plant must follow the set-point signal of the TSO within 1 min. When the solar power plant operates at an active power output below its rated capacity, it shall be able to be operated in every possible operating point in the P-Q capability chart for plant size MSSP as shown in **Figure 16** and LSSP as shown in **Figure 17**. It should be noted that for LSSPs, even at zero active power output, reactive power injection at the PCC shall fully correspond to the P-Q capability chart taking into account the power requirements of auxiliary services, transformers' losses, and solar plant cabling.

The maximum values of the capacitive and inductive reactive power in **Figures 16** and **17** are calculated from the nominal generation capacity of the solar power plant and the power factor limit of 0.95 leading and lagging. Using capacitors and/or reactors to meet the requirements of the P-Q chart at the PCC is acceptable.

4.12 Low fault ride through (LVRT)

The SEGCC stipulates that, in case of a grid fault, the grid-connected solar power plant has to remain connected to the grid when the positive-sequence voltage at the PCC is above the curve shown in **Figure 18**. This defines the ability of the solar power plant to ride through the grid fault without disconnection from the grid. If all line-to-line voltages are below the curve shown in **Figure 18**, the solar power plant shall disconnect from the grid.

During this temporary voltage sag, the solar power plant must satisfy the following reactive power (or reactive current) requirement: in the case of a three-phase fault, the solar power plant must be able to inject reactive current in accordance with the curve shown in **Figure 19**, and satisfying Eqs. (19) and (20) for the time period of 250 ms started at the beginning of the fault and continue until clearing the fault.

Figure 19 shows the minimum reactive current required for the solar power plant during the fault. It is represented as the ratio of the reactive current to the nominal plant reactive current against the voltage drop which is represented as the ratio of the actual voltage to the nominal voltage at the PCC. All currents and voltages are in pu.

The following Eqs. (19) and (20) describe the required injected current during the fault:

$$\frac{\Delta I_B}{I_N} = k \times \frac{\Delta U_r}{U_N} \tag{19}$$

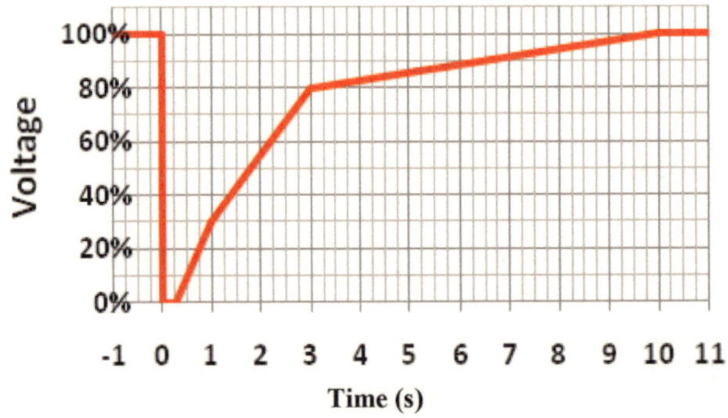

Figure 18.
Low voltage ride-through curve of solar plants.

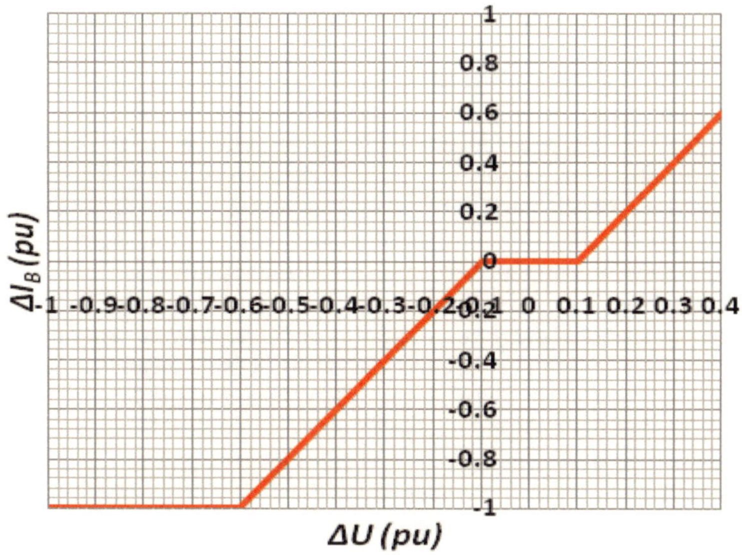

Figure 19.
Requirement of reactive current injection during the fault (k = 2).

$$\Delta U = U - U_O \tag{20}$$

If $\Delta U \geq 0.1$, then $\Delta U_r = \Delta U - 0.1$
If $-0.1 < \Delta U < 0.1$, then $\Delta U_r = 0$
If $\Delta U \leq -0.1$, then $\Delta U_r = \Delta U + 0.1$
If $\Delta U \leq -0.6$, then $\Delta I_B = -1$ pu

where U_N = the rated voltage; I_N = the rated current; U = the voltage during the fault; ΔI_B = the required reactive current change during the fault; U_0 = the voltage pre the fault; ΔU_r = the related change in the voltage during the fault.

In Eq. (19), the factor k shall be adjustable within the range of 0–4. In the case of unsymmetrical faults, it is not permitted to feed reactive currents to the grid during a fault which will cause rise to voltages higher than 110% of the nominal voltage at the PCC in the non-faulty phases. After fault clearance, the active power output from the solar power plant must reach the same value as that of pre-fault value within a period of 10 s after clearing the fault, and the reactive power consumption of the solar power plant must be less than or equal to the reactive power consumption before occurrence of the fault.

5. Comparison of solar energy grid connection codes

Solar energy grid connection codes may be issued as national standards in various countries or by transmission and distribution system operators [22]. These solar energy grid connection codes may be included in the relevant codes or issued separately as a complementary part. For example, the German Association of Energy and Water Industries issued new grid codes for integration of generating power plants to medium-voltage networks. Directives have been released in Germany for connecting electric generation power plants to medium-voltage and low-voltage grids [3]. The directives were based on the results of developing the German Grid Code for integrating renewable power plants into the high-voltage electricity grid [23]. The scope of the directives includes wind power plants, hydroelectric plants, PV solar generating systems, and combined heat and power plants.

In the UK, the Operations Directorate of Energy Networks Association has issued the Engineering Recommendation G83 [24] titled "Recommendations for connecting small-scale type tested embedded generators (up to 16 A/phase, i.e., 11.04 kW three-phase) in parallel with LV distribution systems." The Engineering Recommendation G59 [25] deals with generating plants greater than 11.04 kW up to 50 kW (three-phase). The rules of these engineering recommendations are applicable to all generation power plants irrespective of the type of electric generator and equipment employed for converting energy source into electricity.

The technical and design criteria required for connecting all types of distributed generation power plant are generally set out in the "Distribution Planning and Connection Code" of the UK distribution code [26] and in the "Connection Conditions Code" of the UK Grid Code [27].

In the USA, code standards, guides, and rules for PV systems are available [28–32]. The IEEE has issued a number of standards for integration of distributed energy resources (DERs) into power grids. The IEEE-1547 Standard series concerns with connecting DERs, including PV systems, among others, to electric power systems. The IEEE-2030 series of standards is issued to help implement communications and information technologies to enhance integration of DER with the grid. The National Electrical Code (NEC) Article 690 addresses safety standards for installing PV systems. Other NEC articles may also be applicable to PV installations. The Underwriters Laboratories (UL) Standard-1741 concerns with DER equipment including inverters, converters, and controllers. Standards and technical requirements for solar equipment, installation, etc. are available as guides for states

Code	Requirements in the code
UK Distribution Code	Be able to control the active power for frequency regulations (installed capacity 50 MW)
Germany grid codes for connecting PV systems to the medium-voltage power grid	Be capable of operation at reduced power output (if PCC rated voltage 10 kV) In above system frequency of 50.2 Hz, all generators have to reduce their output power with a gradient of 40%/Hz of the instantaneous available power The output power of the generator is only allowed to increase again as soon as the frequency reduces below 50.05 Hz
CAISO, USA	It is required that the solar plant be capable of providing a frequency response with 5 and 3% droop settings through its governor-like control loop. The definition of the PV plant droop control is the same as that of conventional generating units: $\frac{1}{\text{Droop}} = \frac{\Delta P/P_{\text{rated}}}{\Delta F/60\text{Hz}}$ The dead band of the droop curve is ±36 mHz
Egyptian Solar Energy Plants Grid Connection Code	For grid frequencies in the range from 50.2 to 51.5 Hz, the solar plant has to reduce active power (installed capacity from 500 to 50 MW) The output power must be reduced by $\Delta P = 0.4 \times PM \times (\Delta f/\text{Hz})$ The output power is allowed to increase again as soon as the frequency is below 50.2 Hz

Table 9.
Comparison of active power and frequency control.

and municipalities [28]. A joint report produced by the North American Electric Reliability Corporation (NERC) and the California Independent System Operator (CAISO) provides information to maintain power system reliability while integrating variable energy resources, mainly wind and PV systems [29]. Large PV power plants are normally connected to the transmission grid [30]. Recently in 2019, the National Renewable Energy Laboratory (NREL) published two useful guide books for DER interconnection including current practices and emerging solutions [31] and permitting guide book for small solar systems [32].

As discussed in detail in previous sections of this book chapter, electricity authorities in Egypt have issued complementary documents to the Grid Code and distribution code for connecting solar systems to grids.

Comparisons of some rules in PV grid connection codes of Germany [1, 3, 22], the UK [1, 22], [24–27], the USA [28–32], and Egypt [7–11], [33] are presented here. The comparisons include power and frequency control rules and reactive power control rules. Detailed comparisons are available in [1, 3, 22].

Code	Requirements in the code
Germany grid codes for connecting PV systems to the medium-voltage power grid	In the event of voltage drop of more than 10% the reactive current contribution of at least 2% of the rated current per percent of the voltage drop, the facility must be capable of feeding the required reactive power within 20 ms
USA: requirements for reactive power control of PV power plants	FERC Order 661-A may be applied to PV power plants, and the required power factor range is ±0.95 measured at the Point of Interconnection (POI). It is also required that the PV power plant be capable of providing sufficient dynamic voltage support to guarantee reliability and safety of the system CAISO reactive power requirement stipulates a voltage operation window for PV power plants to provide reactive power at 0.95 pf lagging when voltage level at the POI is within 0.95–1 pu. Also, the PV plant should be able to absorb reactive power at 0.95 pf leading when voltage level at the POI is within the range of 1–1.05 pu
Egyptian Solar Energy Plants Grid Connection Code	For three-phase faults, the solar power plant must inject reactive current for a time period of 250 ms after the beginning of the fault until fault clearance For unsymmetrical faults, it is not permissible that during the duration of the fault, reactive currents be fed into the grid which will give rise to voltages higher than 110% nominal voltage in non-faulty phases at the grid connection point Reactive power of the solar power plant must be equal to or below the consumption of reactive power before the fault
Egyptian Technical Requirements for Connecting Small-Scale PV (ssPV) Systems to Low-Voltage Distribution Networks	"Power factor: The ssPV shall not inject reactive power into the utility network, while the drain of reactive power shall be limited to a power factor of 0.9. This limit applies unless otherwise agreed upon with the utility." The ssPV consumes reactive power

Table 10.
Comparison of reactive power control.

5.1 Active power and frequency control

The main reason for the active power control is to ensure a stable frequency. **Table 9** summarizes the comparison between active power and frequency control rules in the relevant PV grid connection codes of the four countries, the UK, Germany, the USA, and Egypt.

5.2 Reactive power control

Consumption and generation of reactive power must be matched in order to maintain a stable system voltage. **Table 10** presents comparison of reactive power control requirements in PV grid connection codes.

6. Conclusions and recommendations

This chapter has explored technical design specifications, criteria, technical terms, and equipment parameters required to connect Medium-Scale and Large-Scale Solar Plants (MSSP and LSSP) to the electricity networks. The specifications, terms, and parameters have been extracted from the connection code of the MSSP and LSSP, Electricity Distribution Code, and Grid Code. Technical background of these specifications has been discussed in detail. Comparisons of some important rules in the PV grid connection codes of the UK, Germany, the USA, and Egypt have been described. The technical specifications and design criteria presented here are of great importance for planning, design, installations, testing, commissioning and operation, and engineers working in the field of connecting MSSP and LSSP systems to the transmission or distribution grids.

It is recommended to refer to the full versions of the concerned codes to comply with detailed grid connection requirements and successful operation of the solar power systems. Academic researchers are advised to follow the requirements of utility codes in performing research works related to integrating solar power plants into grids.

Appendix: standards of solar plant components

In the stages of designing, manufacturing, and installation of the solar power plant components, relevant international standards must be satisfied. As an example in Egypt, various IEC standards used for these purposes are listed in **Table 11**. All components shall meet the ranges and the operational requirements stipulated in the MSSP and LSSP solar plant connection codes. The solar power plant should be equipped with a synchronizing unit with a proper phase-locked loop to keep the inverter synchronized with the grid to deliver the right amount of power within permissible operational frequency and voltage variations. The rating and short-circuit duties of the switchgear shall comply with the Grid Code requirements. The power transformer efficiency shall be greater than or equal to 96%.

To enable visibility and control, the solar power plant shall be equipped with monitoring and security facilities having remote access communications means. The remote monitoring and controlling, telecommunications equipment, and the communication links shall comply with the requirements of the Grid Code and the distribution code as requirements of relevant case. The SEGCC contains details of specifications of real-time data, measuring, monitoring, and control equipment. The measurements include active power (kW), reactive power (kVAr), active

Solar plant components	IEC standards
Power transformer	IEC Standard 60076 IEC Standard 60085 for electrical insulation and IEC Standard 60214 for tap changer
AC switchgear	IEC Standard 62271
Inverter	IEC Standard 62109-2 IEC Standard 62116
Cabling and accessories in the site	IEC Standard 60227 series for LV (below 1 kV) IEC Standard 60502 series for HV installations
All relevant components	IEC Standard 60068-2 series for basic environmental tests, at least for IEC Standard 60068-2/1 cold, /2 dry, /14 change of temperature, and /30 damp heat
Site implementation	IEC Standard 60,364 series

Table 11.
IEC standards for components of solar power plants in Egypt.

energy (kWh), reactive energy (kVArh), voltages, currents, frequency, solar irradiance, temperature, and voltage and current harmonic distortions (THDv and THDi). The solar power plant shall provide all status signals, including transformer tap position, circuit breakers, disconnectors and earth switches, telecommunication alarms, protection signals at the grid side, inverter, etc. Also, set points of active power, reactive power, or power factor shall be indicated.

Technology solutions which shall be implemented in measuring, monitoring, and control of the solar power plants are described in detail in the SEGCC. The grid protection settings in the solar plants must comply with the requirements stipulated in the SEGCC, unless otherwise agreed with the transmission system operator. At the PCC, the grid protections shall be in compliance with the protection code of the Grid Code [9].

Author details

Omar H. Abdalla[1]* and Azza A.A. Mostafa[2]

1 Helwan University, Cairo, Egypt

2 South Cairo Electricity Distribution Company, Cairo, Egypt

*Address all correspondence to: ohabdalla@ieee.org

References

[1] Craciun BI, Kerekes T, Sera D, Teodorescu R. Overview of recent grid codes for PV power integration. In: 13th International Optimization of Electrical and Electronic Equipment, OPTIM; 24-26 May 2012; Brasov, Romania. pp. 959-965

[2] Larsen C, Brooks B, Starrs T. Connecting to the Grid—A Guide to PV Interconnection Issues. 3rd ed. USA: Interstate Renewable Energy Council, IREC, North Carolina Solar Center; 2000. pp. 1-37

[3] Troester E. New German codes for connecting PV systems to medium voltage power grid. In: 2nd International Workshop on Concentrating Photovoltaic Power Plants: Optical Design, Production, Grid Connection; 9-10 March 2009; Darmstadt, Germany. pp. 1-4

[4] Tweedie A, Doris E. Comparing Germany's and California's Interconnection Processes for PV Systems. Colorado, USA: National Renewable Energy Laboratory, NREL; 2012. pp. 1-60. Available from: http://www.nrel.gov/docs/fy11osti/51814.pdf

[5] IET Standards. Code of Practice for Grid Connected Solar Photovoltaic Systems. UK: IET; 2014. pp. 1-120

[6] Grid connection code for renewable power plants (RPPs). In: Connected to the Electricity Transmission System (TS) or the Distribution System (DS) in South Africa, Version 2.6. South Africa: National Energy Regulator of South Africa, NERSA; 2012. pp. 1-61

[7] EgyptEra. Solar Energy Plants Grid Connection Code—In addition to the Egyptian Transmission Grid Code and The Egyptian Distribution Network. Cairo, Egypt: Egyptian Electric Utility and Consumer Protection Regulatory Authority; 2017. Available from: http://www.egyptera.org

[8] EgyptEra: "Electricity Distribution Code". Cairo, Egypt: Egyptian Electric Utility and Consumer Protection Regulatory Authority; 2017. Available from: http://www.egyptera.org

[9] EETC: "Transmission Grid Code". Cairo, Egypt: Egyptian Electricity Transmission Company, EETC; Available from: http://www.eetc.net.eg/grid_code.html

[10] EgyptEra: "Technical Requirements for Connecting Small Scale PV (ssPV) Systems to Low Voltage Distribution Networks". Cairo, Egypt: Egyptian Electric Utility and Consumer Protection Regulatory Authority; 2014. Available from: http://www.eg yptera.org

[11] Abdalla OH. Technical design specifications and criteria for integrating PV systems into distribution networks in Egypt. In: Keynote Lecture, KL-REN-5, The 1st FUE International Conference on New Energy & Environmental Engineering; 11-13 April 2016; Cairo, Egypt

[12] The International Energy Agency. The IEA Photovoltaic Power Systems Programme (PVPS). Available from: http://www.iea-pvps.org/

[13] New and Renewable. Energy Authority (NERA), Egypt: Available from: http://www.nrea.gov.eg/

[14] https://www.pv-tech.org/news/ib-vogt-breaks-ground-on-166.5mw-solar-project-in-benban-egypt

[15] IBM Research (NYSE: IBM). Available from: http://www-03.ibm.com/press/us/en/pressrelease/40912.wss

[16] http://energy.gov/articles/celebrating-completion-worlds-largest-concentrating-solar-power-plant

[17] https://gizmodo.com/morocco-switches-on-first-phase-of-the-world s-largest-s-1757281810

[18] Schlaich J, Bergemann R, Schiel W, Weinrebe G. Design of commercial solar updraft tower systems-utilization of solar induced convective flows for power generation. Journal of Solar Energy Engineering. 2005;**127**: 117-124

[19] IEC. White Paper. Grid Integration of Large-Capacity Renewable Energy Sources and Use of Large-Capacity Electrical Energy Storage. International Electrotechnical Commission; 2012. Available from: http://www.iec.ch/whitepaper/gridintegration/?ref=extfooter

[20] EgyptEra: "Wind Farm Grid Connection Code—In addition to the Egyptian Transmission Grid Code (ETGC)". Cairo, Egypt: Egyptian Electric Utility and Consumer Protection Regulatory Authority; 2014. Available from: http://www.egyptera.org

[21] Pillay P, Manyage M. Definition of voltage unbalance. IEEE Power Engineering Review. 2001:50-51

[22] Zheng Q, Li J, Ai X, Wen J, Fang J. Overview of grid codes for photovoltaic integration. In: 2017 IEEE Conference on Energy Internet and Energy System Integration (EI2); Beijing: 2017. pp. 1-6. DOI: 10.1109/EI2.2017.8245501

[23] Transmission Code 2007 Network and System Rules of the German Transmission System Operators

[24] Operations Directorate of Energy Networks Association: Engineering Recommendation G83: Recommendations for connecting small-scale type tested embedded generators (up to 16 A/phase) in parallel with LV distribution systems, Issue 2. 2012. Available from: www.energ ynetworks.org

[25] Operations Directorate of Energy Networks Association: Engineering Recommendation G59: Recommendations for the connection of generating plant to the distribution systems of licensed distribution network operators, Issue 3, Amendment 3. 2018. Available from: www.energynetworks.org

[26] The Distribution Code of Licensed Distribution Network Operators of Great Britain, Issue 29. 2018

[27] National Grid Electricity Transmission plc: The Grid Code, Issue 4 Revision 2. 2010

[28] Beren Argetsinger, and Benjamin Inskeep. Standards and Requirements for Solar Equipment, Installation, and Licensing and Certification—A Guide for States and Municipalities. Clean Energy States Alliance; 2017. Available from: www.cesa.org

[29] NERC and CAISO. Maintaining Bulk Power System Reliability While Integrating Variable Energy Resources —CAISO Approach. A joint report produced by: The North American Electric Reliability Corporation and the California Independent System Operator Corporation; 2013. Available from: https://www.nerc.com/pa/RAPA/ra/Reliability%20Assessments%20DL/NERC-CAISO_VG_Assessment_Final.pdf

[30] Loutan C et al. Demonstration of Essential Reliability Services by a 300-MW Solar Photovoltaic Power Plant. National Renewable Energy Laboratory (NREL); 2017. Available from: https://www.nrel.gov/docs/fy17osti/67799.pdf

[31] Peterson Z et al. An Overview of Distributed Energy Resource (DER) Interconnection: Current Practices and Emerging Solutions. Golden, CO: National Renewable Energy Laboratory; NREL/TP-6A20-72102. 2019. pp. 1-84. Available from: https://www.nrel.gov/docs/fy19osti/72102.pdf

[32] Newsom GC et al. California Solar Permitting Guidebook Improving Permit Review and Approval for Small Solar Systems. Updated Fourth Edition. California, USA: Solar Permitting Task Force, Governor's Office of Planning and Research; 2019. pp. 1-114. Available from: opr.ca.gov/docs/20190226-Solar_Permitting_Guidebook_4th_Edition.pdf

[33] Abdalla OH. Technical requirements for connecting medium and large solar power plants to electricity networks in Egypt. Journal of the Egyptian Society of Engineers. 2018;**57**(1):25-36

Chapter 3

Energy Return on Investment Analysis of a Solar Photovoltaic System

Harpreet Kaur and Inderpreet Kaur

Abstract

The consumption of petroleum product assets on an overall premise has required an earnest look for elective vitality sources to get together the present-day request. The world likewise faces the double difficulties of petroleum derivative exhaustion and CO_2 discharges plus the fundamental competitor for confronting these difficulties. However, safe and economic concepts for CCS have not been proven, nuclear suffer from high cost, radioactive waste management, fuel availability, and nuclear weapon proliferation issues, and renewable, other than hydropower, have been limited by resource limits, high cost, and intermittency problems. In any case, the later intense cost decreases in the creation of photovoltaics (PV) which makes ready for empowering sun-based innovations to end up cost focused on petroleum energy generation. The target of present work is to evaluate the capability of sun oriented solar power at Chandigarh University, which lastly built up a framework depending on the potential estimations for a picked region of 1050.1416 m^2. At the end, cost estimation of SPV is determined to indicate whether it is monetarily practical or not.

Keywords: energy, petroleum, sustainable, irregularity, generation, estimation

1. Introduction

Power assumes a critical function in everyday life movements. The level of advancement and progress of the nation is estimated via evaluation of an individual's power use pattern. Power requirement is escalating stepwise with increasing populace, modernization, and commercialization. The global conventional energy source delivery, viz. coal, oil, and gaseous petrol, will consequently be drained in the next 100 years [1]. The speed of energy employment is intensifying; supply is exhausting bringing about inflammation and energy deficiency. It is classified as "power crisis." Hence, possible or inexhaustible wellsprings of vitality have to be created to fulfill future energy necessities. In the previous 200 years, the vitality framework dependent on coal, oil, flammable gas, and other nonrenewable energy sources has extraordinarily advanced the improvement of human culture. However, not only the material life and the spiritual life are increasing, but also serious consequences also brought from the large scale use of fossil fuels are increasing, depleting the resources and deteriorating the environment [2].

It also includes political and economic disputes of a number of nations. Frequently growing anxiety about "global warming" and exhaustion of oil has motivated various nations worldwide to implement novel power techniques to fulfill the power requirement and to conserve our atmosphere [3, 4]. In order to protect the environment and for sustainable growth, the importance of sustainable energy cannot be overemphasized. It leads to speed up the study and progress of sustainable energy techniques particularly PV applications because of its quiet operation, long lifetime, and little repair. PV offer clients the capability to produce electricity in a dirt-free, silent, and trustworthy way. PV systems consist of solar cells, gadgets that convert light energy straight into power [5]. The photovoltaic (PV) system has practiced a rapid growth over the last decade and is expected to accelerate in the next 10–20 years. Recently, a PV system has broadly received a lot of concentration due to many significant advantages such as unlimited accessibility of key energy sources and no polluted emissions [6]. The quantity of PV generation is rising quickly both in size and complexity. As a consequence, the cost of PV systems is constantly declining. It has been noticed that nowadays more and more photovoltaic systems are being installed. Energy classification is shown in **Table 1** [7].

Energy source type	Definition	Example
Primary energy sources	These are the resources which can be legitimately established in nature or put away in nature and can be separated. Accessible in crude from which they should be prepared first for use.	Stored nuclear energy from radioactive material; direct—coal, oil, nature gas, and biomass.
Secondary energy sources	Optional vitality resources are gotten from essential sources in the form of either last fuel or vitality supply. Inclusion of innovative procedures in this change in the middle of makes drop in essential vitality in transit purchasers.	Gasoline, petrol, steam energy from coal, etc.
Waste energy resources	It is possible to reuse waste energy liberated in the process of utilization of primary and secondary energy resources. Practically it is achieved by combined heat and power which is more popular as cogeneration.	Energy extracted from cooling systems in power plants.
Renewable (nonconventional)	This is the energy acquired from never ending sources of energy available in nature. The main feature of this is that it can be extracted without causing pollution.	Solar power, wind energy, geothermal energy, tidal energy, and biomass.
Nonrenewable energy Conventional	Nonrenewable energy is the energy obtained from the conventional fuels which are exhaustible today or tomorrow with time.	Coal, oil, gas, hydropower, diesel power.
Commercial	This is the energy accessible from market at certain price. These are the cardinal source for industrialized countries as its basic need for industries, commercial transport, and agricultural sectors.	Electricity, lignite, coal, oil, natural, gas, etc.
Noncommercial	These sources are not available in the market unlike previous type for a price. Instead, these are traditionally gathered. Also termed as traditional fuel and mostly shrugged off in energy accounting.	Firewood, cattle, dung sugarcane crush, solar and thermal water heating, etc.

Table 1.
Classification of energy sources [7].

1.1 Present energy status

The current energy status is discussed in the form of world, India, and Punjab.

1.1.1 World energy scenario

World energy demand has been growing exponentially in this century as shown in **Figure 1**. By means of fossil fuels like coal, wood, oil, or gas, it was possible for humankind to set up a society for colder climates. Because of the expanding requests for solar energy, a higher portability and a bigger total populace, vitality utilization is immensely raised in the course of the most recent 150 years. The International Energy Agency (IEA) estimates that global main energy between now and 2030 will increase by 1.5% per year [7].

1.1.2 Energy status in India

Before the finish of the year 2019, India had an energy generation limit of around 155 Giga Watt. But still, around 17% (450 million) of towns in India are not electrified [8–10]. With a developing financial system, the interest for energy is developing at 6% consistently and the top burden request is relied upon to achieve 225 GW before the end of the year 2022.

The Indian power division is extremely reliant on coal for energy requirement which is about 53% of the overall capacity. As per the present scenario, coal utilization by power division is probably to arrive at 200 Million Metric Tonnes by 2013. As per the Coal Ministry, the accessible coal resources are predicted to remain for the next 40–45 years. Around 11% of the entire power is received from oil and gas [10–13]. As shown in **Figure 2**, in India, the power division is the major user of oil and gas other than automobiles and industry.

In India, individual energy consumption is extremely lesser as compared to the world level. Indeed, even with such a low individual capita utilization, the power shortfall is around 11% in whole requirement and a shortage of over 12% in crest load request [13]. This obviously means the accessible fuel is not adequate to satisfy the rising need for vitality in India.

1.1.3 Solar energy situation in Punjab

Energy preservation is a standout among the main significant focuses today and Punjab is blessed with tremendous capability of sun powered vitality with more than 300 days of daylight per year and renewable energy source is being adequately developed by PEDA. As the state is endowed with vast potential of solar energy

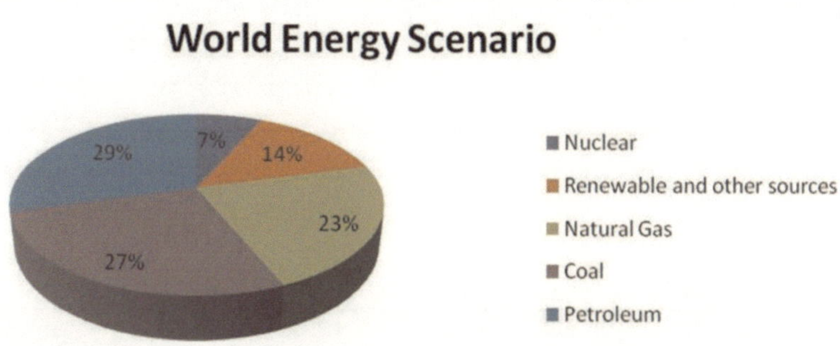

Figure 1.
Energy generation capacity in the world [7].

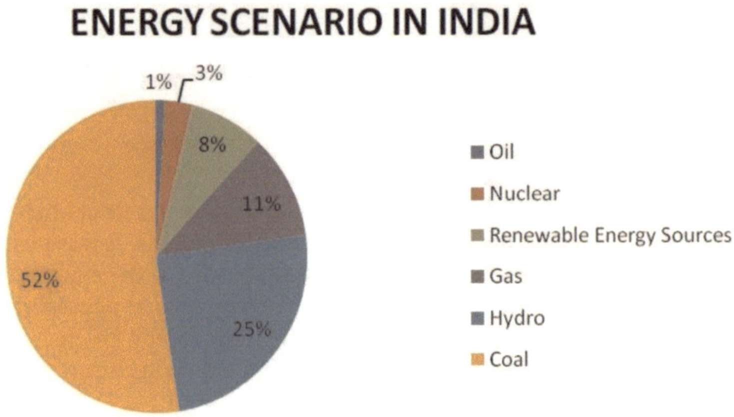

Figure 2.
Power generation capacity in India [7].

estimated at 4 -7 KWH/Sq.mtr of solar insolation levels, the Government is also keen to tap this resource for strengthening power infrastructure in the State by setting up solar energy-based power projects so as to save the depleting resources for our future generation and to control global warming, fast depleting conventional sources of energy and resultant increased environmental pollution. PEDA with its untiring endeavors has effectively charged numerous ventures in various conditions of Punjab [14]. Under the rooftop programme, SPV power projects are being set up at various important Government, Institutional and Religious buildings, namely Punjab Raj Bhawan, Golden Temple, Wagah Border, Punjab Agricultural University, Ludhiana, and Pushpa Gujral Science City, Kapurthala.

1.2 Solar energy in India

India is a region which receives a good amount of solar heat flux. **Table 2** gives mean daily solar radiations at some places in India.

During the last few years, more than 50 small solar power plants have been installed in villages in Uttar Pradesh and other states. The size of these plants varies from 2 to 10 KW. These plants are supplying power to these remote villages where power from grid is not available. In addition, more than 40,000 solar powered street lighting systems have been installed in different states and more than 1000 water pumping systems have also been installed. It is proposed to use more than

Place	Mean daily solar radiations KWH per meter square
Port Blair	4.3
Madras	5.6
Poona	5.7
Jodhpur	6
Delhi	5.4
Shilling	4.4
Western Rajasthan	7.4

Table 2.
Mean daily solar radiations at different places in India [14].

20,000 photovoltaic power packs (each about 70–90 W range) for rural telephone system in the near future. In all, more than 50,000 photovoltaic systems are being used at different places [15–17].

In addition to above, solar water heating system, solar cookers, solar air heaters, solar driers for food grains, etc. are being used on a very large scale [16]. It is expected that the total solar power utilization in India would be around 10,000 MW by the year 2020.

2. Problem formulation

2.1 Motivation

The interest of power has expanded and that much interest cannot be gained by the ordinary power plants, and furthermore, in the creating nations, similar to India, there is a serious issue of lack of intensity [18, 19]. The predominance of the power supply in a couple of spots is described by vast voltage and recurrence variances, arranged and unconstrained power cuts, and burden top limitations. This has prompted quick utilization of sun-oriented capacity to meet the fundamental burden [20]. This framework offers a superior productivity, adaptability of arranging, and ecological advantages contrasted with power frameworks [20, 21]. In this manner, my venture work centers around the independent PV framework joining sun-based PV and battery back-up in the scholastic grounds in light of the fact that a large portion of the activity of scholarly grounds happen in the day time, which is in synchronous with the accessibility of daylight.

2.2 Need and significance of work

Indian power framework is constantly overburdened because of hole between the free market activity, and in this manner, control in the optional conveyance framework is not constantly accessible. Vitality is a critical contribution for monetary improvement [21]. Vitality is required for financial development, for improving the personal satisfaction, and for expanding open doors for advancement. Since expendable vitality sources in the nation are restricted, there is an earnest need to concentrate on the improvement of sustainable power sources and utilization of vitality proficient advances [21, 22]. Notwithstanding nonsolid framework, there is not really a constrained interest of sun-powered photovoltaic frameworks in the business part of a creative nation like India, the primary purpose behind this being the absence of information and inaccessibility of monetarily appealing or reasonable sun-based PV framework [23]. The estimations with regard to the independent power age framework with capacity bank at Chandigarh University, Gharuan, Mohali will be proposed here and there is sufficient energy from the sun that can be provided from the PV-cluster framework. At whatever point there is abundance supply from the sustainable power source, the vitality stockpiling bank stores vitality which will be utilized on the occasion when there are inadequate supplies from the sustainable power source.

2.3 Purpose of the study

- The target of this work is to analyze the specialized achievability and efficient practicality of the PV system with battery reinforcement framework with the available Chandigarh University infrastructure

2.4 Methodology

- Step I: basic load will be determined.
- Step II: total day time energy required will be determined.
- Step III: night time total energy and power required will be determined.
- Step IV: payback period will be calculated

3. Present work

Photovoltaics offer the ability to deliver control in an ideal, calming, and reliable way. Photovoltaic systems contain photovoltaic cells, contraptions that convert light vitality directly into power. Since the wellspring of light is typically the sun, they are consistently called sun powered cells [21, 24]. The word photovoltaic starts from "photo" implying light and "voltaic" which insinuates conveying power. Along these lines, the photovoltaic methodology is "conveying power explicitly from sunlight." Photovoltaics are normally abbreviated as PV.

3.1 Solar energy

The vitality from the sun is given as radiations. The imperativeness is made in the sun's inside through the mix of hydrogen particles into helium. By and by, in light of greater detachment of sun from the Earth, only a little portion of sun's radiations accomplishes the Earth's surface [21]. The power of daylight-based radiation accomplishing the Earth's surface is around 1369 watts for each square meter [w/m^2]. This is known as the "solar constant". The total solar radiation intercepted by earth's surface can be calculated by multiplying solar constant with cross-sectional area of the earth [26]. In order to calculate the solar radiation received, on average per square meter of earth's surface, we divide the above multiplied result by the surface area of the earth. Thus, the average solar radiations per square meter of earth's surface is given by Eq. (1).

$$R = \frac{S\pi r^2}{4\pi r^2} = \frac{1369}{4} \approx 342 \frac{W}{m^2} \tag{1}$$

where S is the solar constant in W/m^2 and r is the radius of Earth.

3.2 Solar energy realized at the Earth's surface

Till now, the effect of Earth's condition is not thought about. The esteem decided above is for the typical sun-fueled radiation drive at the outer areas of the Earth's atmosphere. So we are charmed to know that the measure of this vitality truly accomplishes the Earth's surface. Nature absorbs around 68 W/m^2 and reflects 77 W/m^2.The radiation achieving the world's surface is 198 W/m^2.The force of sun-oriented radiation additionally relies upon the time and the topographical positions [26]. From **Figure 3**, we can see that each square meter of the upper areas of air gets 342×67 W/m^2 of vitality consumed by the climate and 77 W/m^2 is reflected as shown in **Figure 3**.

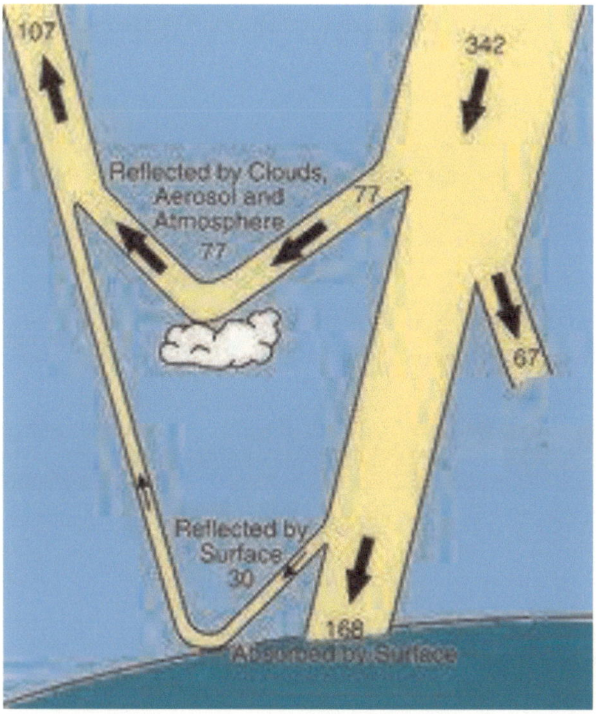

Figure 3.
Distribution of solar radiation [2].

Around 168 W/m^2 vitality lands at the world's surface and 30 W/m^2 is reflected back to space. The energy that reaches the earth surface is around 3.2 EJ/y if we are able to harvest even a small fraction of the available energy at the earth surface we could solve our energy problems. The energy reaching the earth surface is around 7000 times the global energy consumptions [27, 28].

3.3 Solar energy band

The proficiency of a PV gadget is subject to the unearthly appropriation of sun-powered radiation. The assessment of PV gadgets is commonly finished with reference to a standard unearthly circulation. There are two standard earthly appropriations characterized by the American Society for Testing and Materials (ASTM), direct ordinary and worldwide AM 1.5. The immediate typical standard relates to sunlight-based radiation that is opposite to a plane straightforwardly confronting the sun [22, 29]. The worldwide relates to the range of the diffuse radiations. Radiations which are thought about the Earth's surface or affected by climatic conditions are called diffuse radiations. To quantify worldwide radiations, an instrument named pyranometer is utilized [29]. This instrument is planned so that it reacts to each wavelength so that we get an exact incentive for all out power in any occurrence range.

The AM initials in **Figure 4** above represent air mass. The air mass in this setting implies the mass of air between a surface and the sun. The length of the way of sunlight-based radiation from the sun through the environment is shown by the number AMx [19]. The more drawn out the way the more is the divergence of light.

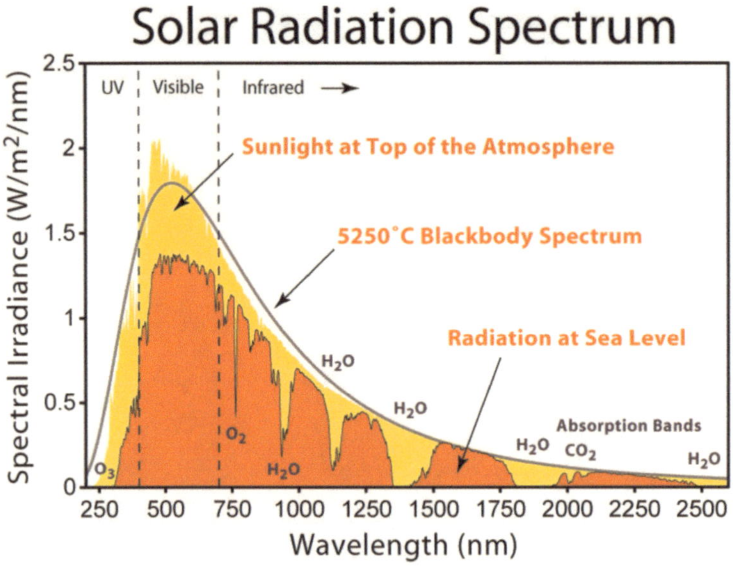

Figure 4.
Solar spectral distribution [29].

3.4 Solar constant

The rate at which solar energy arrives at the top of the atmosphere is called solar constant. This is the amount of energy received in unit area perpendicular to the sun's direction at the mean distance of the Earth from the sun. The rays' focus on the outside of the sun is about 6.33×107 W/m^2. In perspective on the way that radiation spreads out as the separation squared, when it goes to the Earth (1.496×1011 m or 1 AU is the normal Earth-sun remove), the brilliant vitality falling on 1 m^2 of surface zone is diminished to 1367 W as portrayed in **Figure 5**. The intensity of radiation leaving the sun is relatively constant. Therefore, the intensity of solar radiation at a distance of 1 AU is called the solar constant Isc.

3.5 System components

The stand-alone power system, described here, basically includes the following main elements.

1. Renewable energy source: PV system

2. Energy storage bank: battery bank

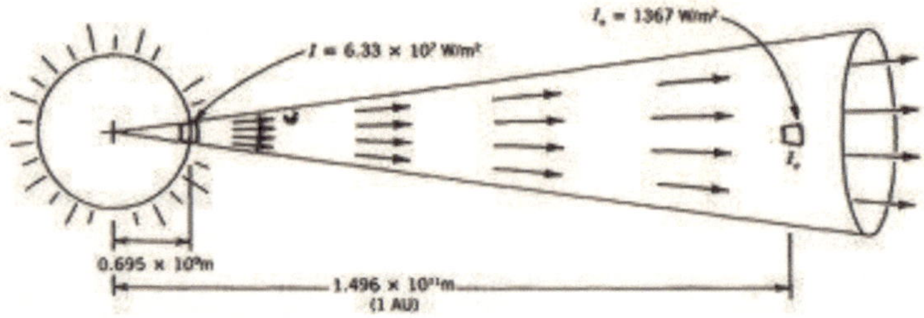

Figure 5.
The distribution of energy from the sun to the Earth [3].

3. AC loads

4. Power electronic devices

The detailed descriptions are given one by one in the following topics.

4. Photovoltaic system

4.1 Description

A photovoltaic structure is a system which uses something like one sun situated sheets to change over sun-arranged imperativeness into power. It contains various parts, including the photovoltaic modules, mechanical and electrical affiliations, and mountings and techniques for coordinating just as modifying the electrical yield. A sun-controlled cell (moreover called a photovoltaic cell) as shown in **Figure 6** is an electrical device that changes over the imperativeness of light (optical essentialness) clearly into power by the photovoltaic effect. It is a sort of photo-electric cell (in that its electrical characteristics—for instance stream, voltage, or deterrent—change when light is incident upon it) which, when introduced to light, can make and support an electric stream without being attached to any external voltage source.

4.2 Semiconductor structure

Semiconductors, for instance, silicon (Si), contain individual atoms fortified together in a conventional, discontinuous structure to outline a game-plan whereby each particle is incorporated by eight electrons. An individual atom contains a center made up of a focal point of protons (determinedly charged particles) and neutrons (particles having no charge) incorporated by electrons. The amount of electrons and protons is comparable, with the true objective that the atom is commonly electrically fair. The electrons enveloping each particle in a semiconductor are a bit of a covalent bond. A covalent bond includes two particles "sharing" a lone electron. Each molecule shapes four covalent bonds with the four incorporating particles. Thus, between each particle and its four including particles, eight electrons are being shared. The structure of a semiconductor is shown in **Figure 7**.

Each line connecting the atoms represents an electron being shared between the two. Two electrons being shared are what form the covalent bond.

Supplanting a silicon particle with an atom that has either three or five valence electrons will along these lines produce either a space with no electron (an opening) or one extra electron that can move more uninhibitedly than the others; this is the

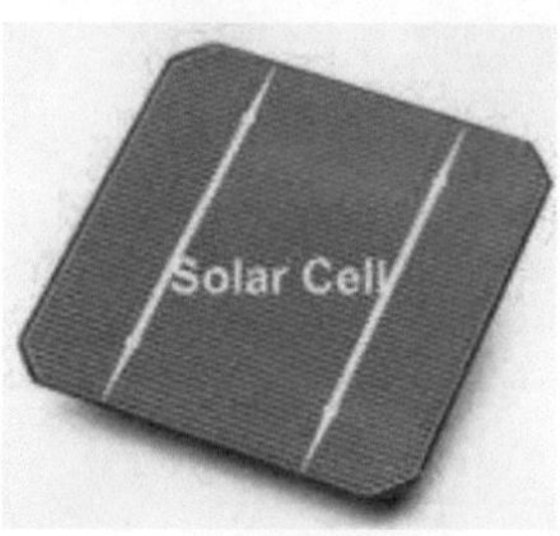

Figure 6.
Solar cell.

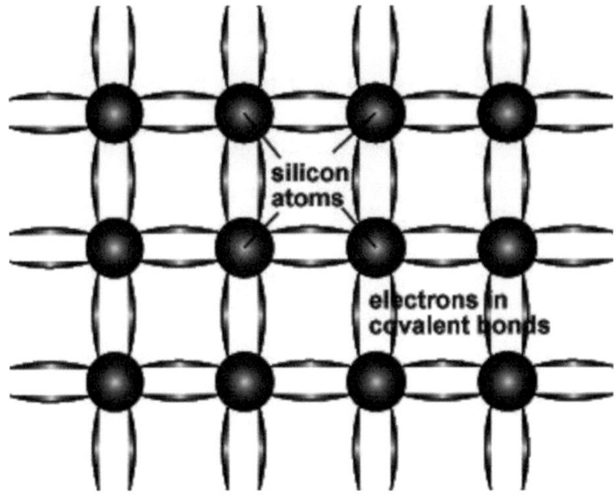

Figure 7.
Schematic representation of covalent bonds in a silicon crystal lattice.

premise of doping. P-type doping, the making of abundance gaps, is accomplished by the consolidation into the silicon of molecules with three valence electrons, regularly boron and n-type doping; the formation of additional electrons is accomplished by fusing a particle with five valence electrons, frequently phosphorus (see **Figure 8**).

When a p-n intersection is made, electrical contacts are made to the front and the back of the cell by vanishing or screen printing metal onto the wafer. The back of the wafer can be totally secured by metal; however, the front just has a network design or slight lines of metal for the occurrence of light photons.

4.3 Conduction in semiconductors

The bond structure of a semiconductor decides the material properties of a semiconductor. One key impact is the vitality level which the electrons can possess

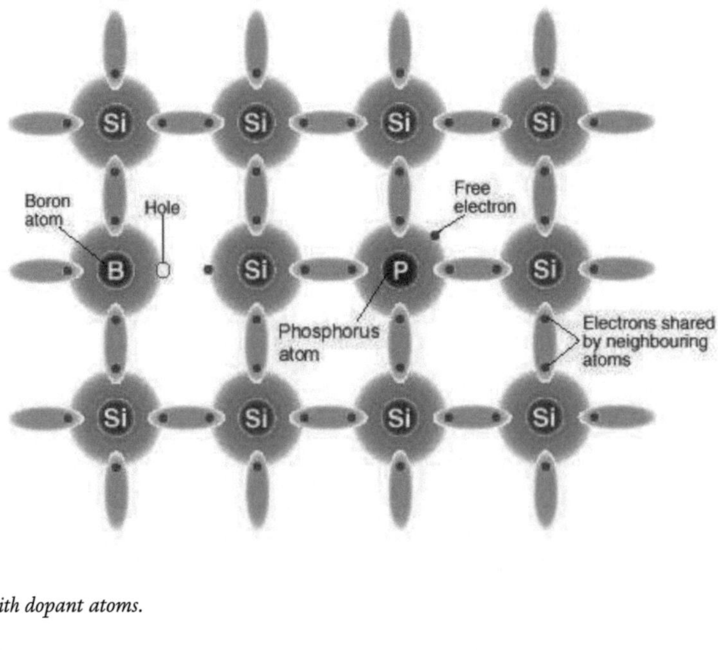

Figure 8.
Si crystal with dopant atoms.

and how they move about the precious stone cross section. The electrons in the covalent bond shaped between every one of the molecules in the cross section structure are held set up by this bond and consequently they are restricted to the locale encompassing the particle. These reinforced electrons cannot move or change vitality, and therefore are not considered "free" and cannot take an interest in current stream, adsorption, or other physical procedures of enthusiasm for sun-based cells. However, only at absolute zero, all electrons are in this "stuck", bonded arrangement. At elevated temperatures, especially at the temperatures where solar cells operate, electrons can gain enough energy to escape from their bonds. At the point when this occurs, the electrons are allowed to move about the gem cross section and partake in conduction. At room temperature, a semiconductor has enough free electrons to enable it to lead to current. At or near supreme zero, a semiconductor carries on like an insulator.

At the point when an electron increases enough vitality to take an interest in conduction (is "free"), it is at a high vitality state. At the point when the electron is bound, and along these lines cannot take an interest in conduction, the electron is at a low vitality state. Along these lines, the nearness of the bond between the two iotas presents two unmistakable vitality states for the electrons. The electron cannot achieve vitality esteems middle of the road to these two dimensions; it is either at a low vitality position in the bond, or it has sufficiently increased vitality to break free and subsequently has a specific least vitality. This base vitality is known as the "band hole" of a semiconductor as shown in **Figure 9**. The band hole of a semiconductor is the base vitality required to energize an electron that is stuck in its bound

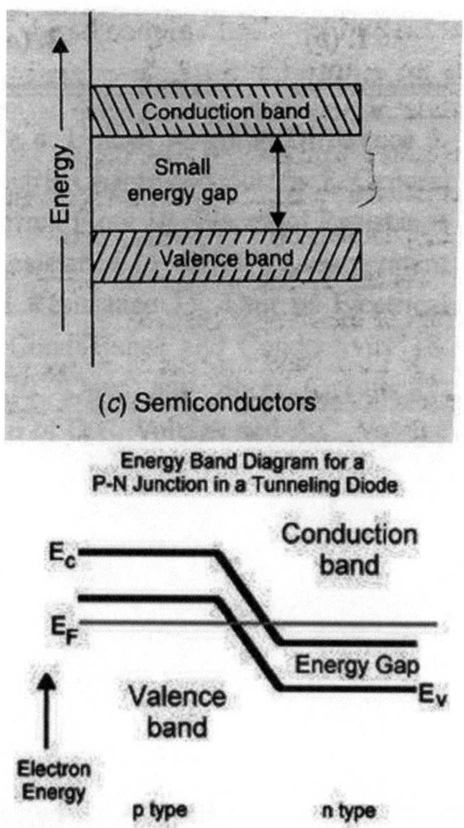

Figure 9.
Energy band diagram.

state into a free state where it can take an interest in conduction. The band structure of a semiconductor gives the vitality of the electrons on the y-pivot and is known as a "band graph." The lower vitality dimension of a semiconductor is known as the "valence band" (EV) and the vitality level at which an electron can be viewed as free is known as the "conduction band" (EC). The band hole (EG) is the hole in vitality between the bound state and the free state, between the valence band and the conduction band. In this manner, the band hole is the base change in vitality required to energize the electron with the goal that it can partake in conduction.

The number and vitality of these free electrons, those electrons taking an interest in conduction, are essential to the task of electronic gadgets. The space deserted by the electrons enables a covalent attach to move starting with one electron then onto the next, in this way giving off an impression of being a positive charge traveling through the precious stone cross section. This vacant space is normally called a "gap," and is like an electron, however with a positive charge. **Figure 10** shows the whole process.

4.4 Types of photovoltaic cells

The use of silicon in the manufacture of photovoltaic cells produces the stereo typical uniform blue-colored PV cell as seen on roof tops and the sides of buildings. Solar cells are made from a wide range of semiconductor materials. Currently, the two main solar technologies are crystalline silicon (silicon wafers) and thin film deposits, which vary from each other in terms of light absorption efficiency, energy conversion efficiency, manufacturing technology, and cost of production. Crystalline silicon is cut from a bulk material. It can be single-crystalline, multicrystalline, and amorphous. These cells are the most common type of photovoltaic cell in use today and are also one of the earliest successful PV devices. There are essentially two types of PV technology, crystalline and thin film. Crystalline can again be broken down into two types.

4.4.1 Crystalline silicon (C-Si)

Crystalline photovoltaic cells are produced using silicon which is first softened, and afterward solidified into throwing of unadulterated silicon. Meager cuts of silicon called wafers are cut from a solitary precious stone of silicon

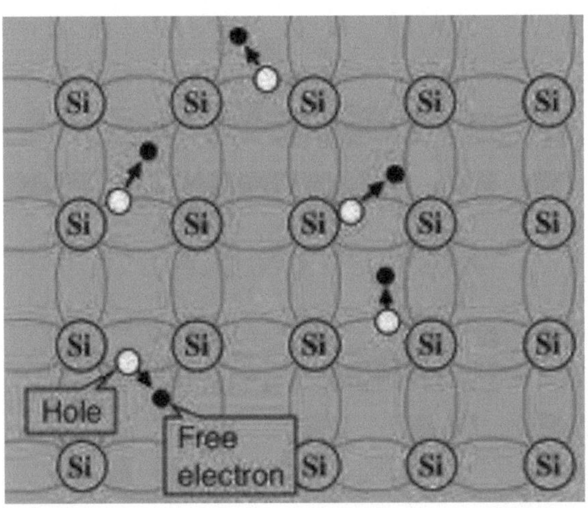

Figure 10.
Generation of holes and electrons.

(monocrystalline) or from a square of silicon gems (poly-crystalline) to make singular cells. The change in productivity for these kinds of photovoltaic cell runs somewhere in the range of 10–20%.

a. **Monocrystalline silicon**: it is a kind of photovoltaic cell material produced from a solitary gem silicon structure which is uniform fit as a fiddle in light of the fact that the whole structure is developed from a similar precious stone. High immaculateness silicon is liquefied in a pot as shown in **Figure 11**. A solitary gem silicon seed is plunged into this liquid silicon and is gradually hauled out from the fluid creating a solitary precious stone ingot. The throwing is then cut into slight wafers or cuts which are then cleaned, doped, covered, interconnected, and collected into modules and exhibits. These kinds of photovoltaic cells are additionally generally utilized in photovoltaic board development. The conversion efficiency for a monocrystalline cell ranges between 15 and 20%. They are highly reliable for outdoor power applications due to their wafer thickness.

b. **Polycrystalline silicon**: these are called multicrystalline silicon and are cast to deliver a silicon ingot. The silicon sub-atomic structure comprises of a few littler gatherings or grains of precious stones, which present limits between them. **Figure 12** demonstrates polycrystalline PV cells. These cells are less vitality effective than monocrystalline silicon PV cells in light of the fact that these limits confine the stream of electrons through them by urging the negative electrons to recombine with the positive openings decreasing the power yield of the cell. The aftereffect of this implies a polycrystalline PV cell which just has a vitality transformation productivity of between 10 and 14%. Nonetheless, these kinds of photovoltaic cells are significantly less costly to create than the proportionate single monocrystalline silicon because of their lower production costs.

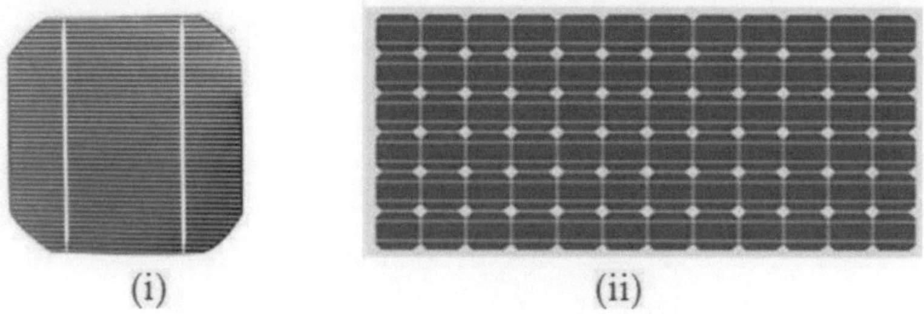

Figure 11.
Monocrystalline cell (i) and panel (ii).

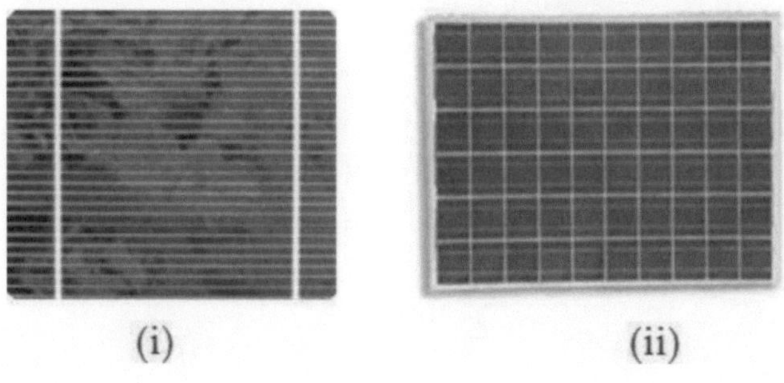

Figure 12.
Polycrystalline cell (i) and panel (ii).

c. **Amorphous silicon**: amorphous silicon is a noncrystalline form of silicon that is widely used in calculators, consumer electronics, and solar garden products that require a small current at a low voltage. Amorphous silicon can be deposited on a variety of low-cost rigid and flexible substrates such as polymers, thin metals, and plastics as well as tinted glass for building integration as shown in **Figure 13**. It has very low conversion efficiency ranging between 7 and 9% when new, degrading down within a few months of exposure to sunlight to less than 5%.

4.4.2 Thin film solar cell

These cells use thin layers of photovoltaic materials deposited onto a substrate as shown in **Figure 14**. This material can be polycrystalline, like cadmium telluride (CdTe), copper indium diselenide (CIS), and thin-film Si, single-crystalline, such as gallium arsenide (GaAs), as well as organic. Crystalline silicon carries on to make up more than 90% of PV panels produced globally. **Table 3** gives comparison of different technologies.

4.5 Photovoltaic array

PV cells are the basic building blocks of PV modules for almost all applications. The one-half volt produced by a single cell is inadequate. Therefore, desired power, voltage, and current can be obtained by connecting individual modules in series and parallel combinations. At the point when modules are fixed together in a solitary mount, they are known as a board and when a few boards are utilized together, they are called an exhibit. At the point when circuits are wired in arrangement, the voltage of each board is included yet the amperage continues as before [11]. When the circuits are wired in parallel, the voltage of each panel remains the same and the

Figure 13.
Amorphous silicon panel.

Figure 14.
Thin film solar cell.

Sr. no	Technology	Efficiency	Cost
1.	Monocrystalline silicon	12.5–15%	High
2.	Polycrystalline silicon	11–14%	Medium
3.	Amorphous silicon (a-Si)	5–7%	Low

Table 3.
Comparison of different solar technologies.

amperage is added. This wiring principle is used to build photovoltaic (PV) modules. Photovoltaic modules can then wired together to create PV arrays.

The PV cells in a module can be wired to any ideal voltage and current. The measure of current created is straightforwardly relative to the cell's size, transformation proficiency, and the force of light. Gatherings of 36 arrangement associated PV cells are bundled together into standard modules that give an ostensible 12 Volt (or 18 Volts @ apex power) as shown in **Figure 15**.

4.6 Basic theory of photovoltaic cell

The PV cell (solar cell) is a light sensitive, two-terminal, semiconducting p-n junction made of a semiconductor material such as silicon. The photovoltaic cells contain one or more p-n junctions. A solar cell has two layers called P-type and N-type and two corresponding electrodes, negative and positive. N-type material is obtained by doping a silicon crystal with N-type impurity. P-type material is obtained by doping a silicon crystal with P-type impurity.

The N-type layer is thin and clear. The P-type layer is wide. At the point when daylight strikes the N-type meager layer, a portion of the light vitality infiltrates up to the P-type layer. The vitality from "photons" in the light waves is bestowed to the particles and iotas in p-n intersection, bringing about freedom of electron-gap sets. Electrons are discharged from the N-type material and holes are made in P-type material. Electrons are negative charges and holes are positive charges. When the external circuit is completed by connecting electrodes to the load, the electrons flow in the closed external circuit from N-type terminal to P-type terminal. Direction of current is from positive terminal to negative terminal in the external circuit. **Figure 3** shows the photovoltaic principle. Within the P-N junction, "electron-hole" pairs are continuously generated during the incidence of the sunlight. Energy from solar rays is captured by the solar cell and is converted directly to electrical energy as shown in **Figure 16**.

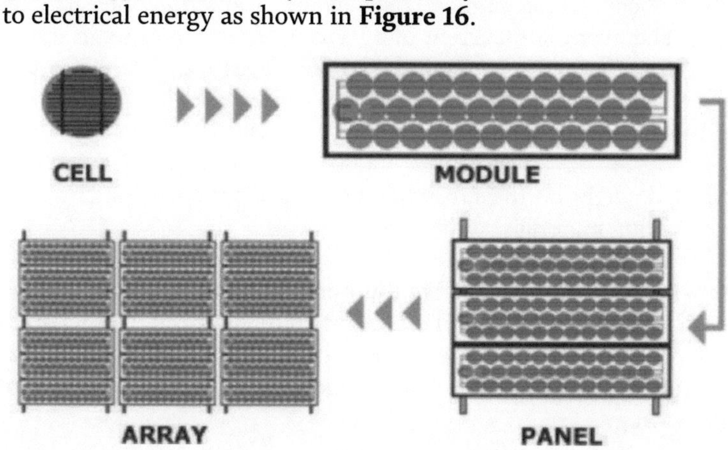

Figure 15.
Photovoltaic hierarchy.

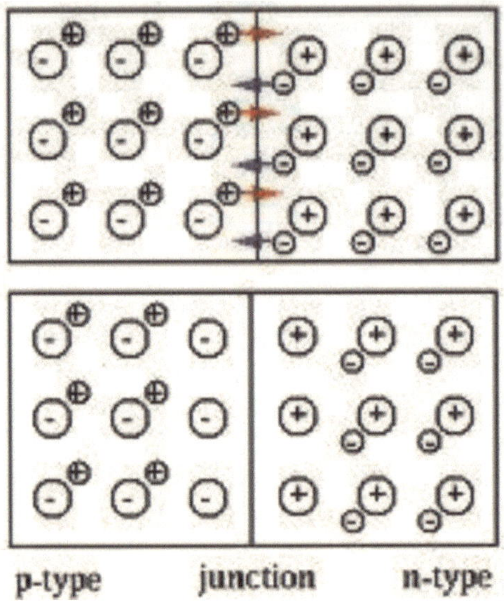

Figure 16.
Structure of P-N junction.

4.6.1 Working of PV cell

Working of a PV cell is based on the basic principle of photovoltaic technology. PV cells contain light engrossing materials (for the most part semiconductors), which are extraordinarily treated to create p-n intersections. A p-n intersection contains two areas of a material with contrasting kinds of conductivity isolated by a semi-conductive intersection. At the point when light falls on the highest point of the cell, it can enter through the intersection and exchange a segment of its vitality to certain electrons, which can cross the intersection into the n-type locale. In the meantime, the integral positive charges called gaps are made in the p-type district where the electrons left. This makes a voltage of 0.5–0.7 Volts over the intersection under open-circuit condition [12]. In the event that an outside circuit is associated with the cell's terminals, the cell will create an immediate current through this circuit. As the outcome, electric power is removed. When the current flows, electrons recombine with holes in the p-type region. As the current increases, the voltage across PV cells drops. **Figure 17** shows the production of electricity directly from sunlight. The maximum current of a PV cell depends on its surface area and intensity of light radiation. In general, the larger the area, the more power can be produced.

The sun-based cell works in three stages:

1. Photons in daylight hit the sun-based board and are consumed by semiconducting materials, for example, silicon.

2. Electrons (contrarily charged) are thumped free from their particles, causing an electric potential contrast. Flow begins coursing through the material to drop the potential and this power is caught. Because of the exceptional organization of sun-powered cells, the electrons are just permitted to move in a solitary bearing.

3. An exhibit of sun-based cells changes over sun-oriented vitality into a usable measure of direct flow (DC) power.

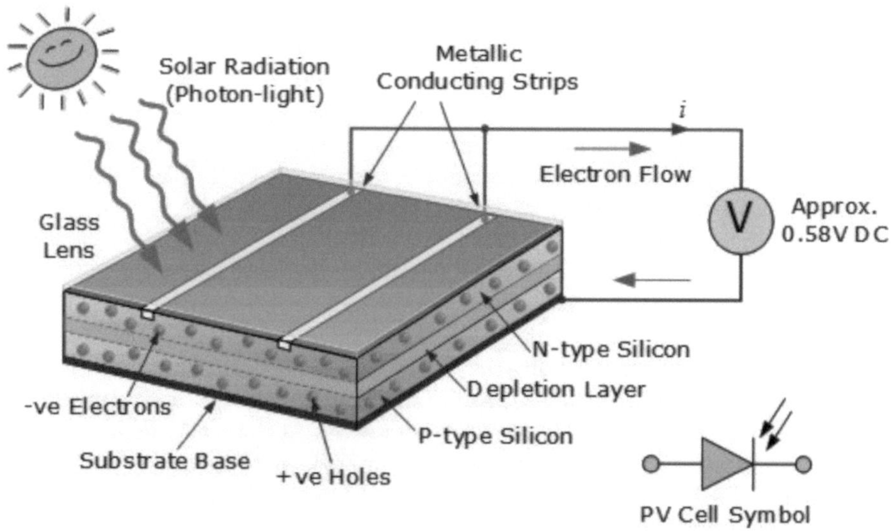

Figure 17.
Production of electricity.

4.6.2 The need for solar cells

The advancement of sun-based cell has been invigorated by:

- The requirement for low upkeep, dependable wellsprings of power appropriate for spots remote from both the principle power network and from individuals; for example, satellites, remote site water siphoning, outback broadcast communications stations, and beacons.
- The requirement for practical power supplies for individuals remote from the principle power matrix; for example, native settlements, outback sheep and steers stations, and some home destinations in framework-associated regions.
- The requirement for nondirtying and quiet wellsprings of power; for example, visitor locales, trains, and campers.
- The requirement for a helpful and adaptable wellspring of little measures of intensity; for example, mini-computers, watches, light meters, and cameras.
- The requirement for inexhaustible and practical power, as a method for decreasing an Earth-wide temperature boost.

4.7 Photovoltaic characteristics

To safeguard similarity with capacity batteries or burdens, it is important to know the electrical qualities of photovoltaic modules. As an update, "I" is the shortening for current, communicated in amps. "V" is utilized for voltage in volts and "R" is utilized for opposition in ohms.

4.7.1 Photovoltaic V-I characteristics

A photovoltaic module will deliver its most extreme current when there is basically no obstruction in the circuit. This would be a short out between its positive and negative terminals. This most extreme current is called hamper, curtailed Isc.

At the point when a module is shortened, the voltage is zero. On the other hand, the most extreme voltage is delivered when there is a break in the circuit. This is called open circuit voltage, condensed Voc. Under this condition, the opposition is interminably high and there is no current, since the circuit is deficient. These two boundaries in burden opposition and the entire scope of conditions in the middle of them are portrayed on a diagram called an I-V (current-voltage) bend. Current, communicated in amps, is on the vertical Y-hub. Voltage, in volts, is on the flat X-pivot as in **Figure 18**.

In the above **Figure 18**, Isc happens on a point on the bend where the voltage is zero. The open circuit voltage happens where the current is zero. The power accessible from a photovoltaic module anytime along the bend is communicated in watts. Watts are determined by duplicating the voltage times the current (Watts = Volts × Amps, or W = VA). At the short out current point, the power yield is zero, since the voltage is zero as shown in **Figure 19**. At the open circuit voltage point, the power yield is likewise zero, yet this time it is on the grounds that the current is zero. There is a point on the "knee" of the bend where the most extreme power yield is found. This point on our model bend is the place the voltage is 17 Volts, and the current is 2.5 Amps. In this manner, the most extreme power in watts is 17 Volts occasions 2.5 Amps, leveling with 42.5. For the maximum output power from a cell, the face of the photovoltaic should be pointed as straight toward the sun as possible. The voltage, current, and power delivered by the solar cell are influenced by

- Conditions of sunlight, intensity, wavelength, angle of incidence, etc. Visible band gives maximum power.
- Conditions of the junction, temperature, termination, etc. It believes there is no shading on the module.

4.7.2 Influence of solar radiation on V-I characteristic of photovoltaic module

Standard daylight conditions on a clear morning are thought to be 1000 Watts of sun powered vitality per square meter (1000 W/m^2). This is now and again called

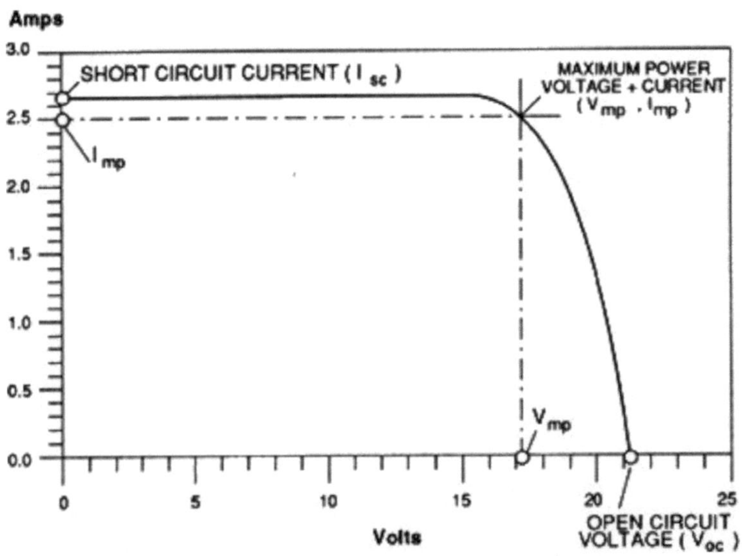

Figure 18.
V-I characteristic curve of photovoltaic module.

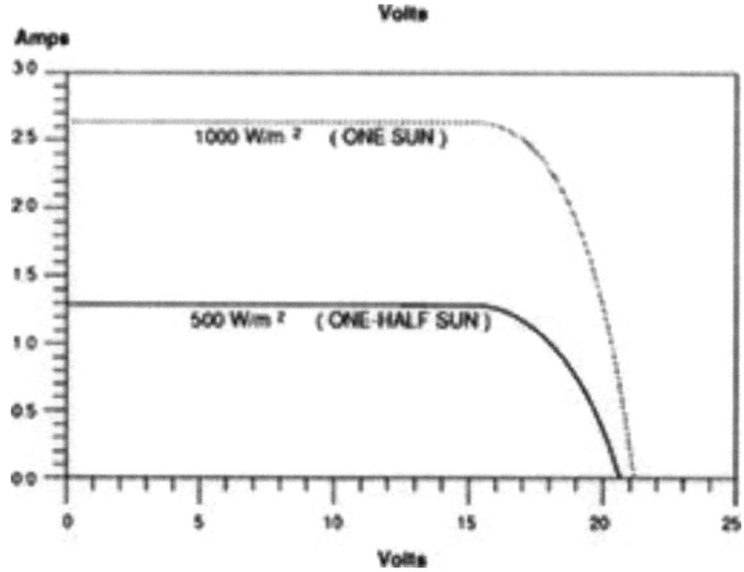

Figure 19.
Change in photovoltaic module voltage and current on change in solar radiation.

"one sun," or a "crest sun" short of what one sun will decrease the present yield of the module by a corresponding sum. For instance, if just one-half sun (500 W/m^2) is accessible, the measure of yield current is generally cut in half for maximum output; the face of the photovoltaic modules should be pointed as straight toward the sun as possible as shown in **Figure 19**.

4.7.3 Influence of temperature on V-I characteristic of photovoltaic module

Module temperature influences the output voltage inversely. High module temperatures will decrease the voltage by 0.04–0.1 Volts for each 1°C increase in temperature (0.04–0.1 V/°C) as shown in **Figure 20**.

That is the reason modules ought to be totally unshaded amid activity. A shadow over a module can nearly stop power generation. Slim film modules are not as influenced by this issue; however, they should in any case be unshaded.

4.8 Photovoltaic system types

PV technology was first applied in space, by providing electricity to satellites. Today, PV systems can be used to power just about anything on Earth. On the basis of working operation PV systems operate in three basic forms.

4.8.1 Standalone PV system

A separate or standalone PV system is comprised of various individual photovoltaic modules (or boards) more often than not of 12 Volts with power yields of somewhere in the range of 50 and 100+ Watts each. These PV modules are then consolidated into a solitary cluster to give the ideal power yield as shown in **Figure 21**. A basic independent PV framework is a programmed close planetary system that produces electrical capacity to charge banks of batteries amid the day for use around evening time when the sun's vitality is inaccessible [11]. An independent little scale PV framework utilizes battery-powered batteries to store the

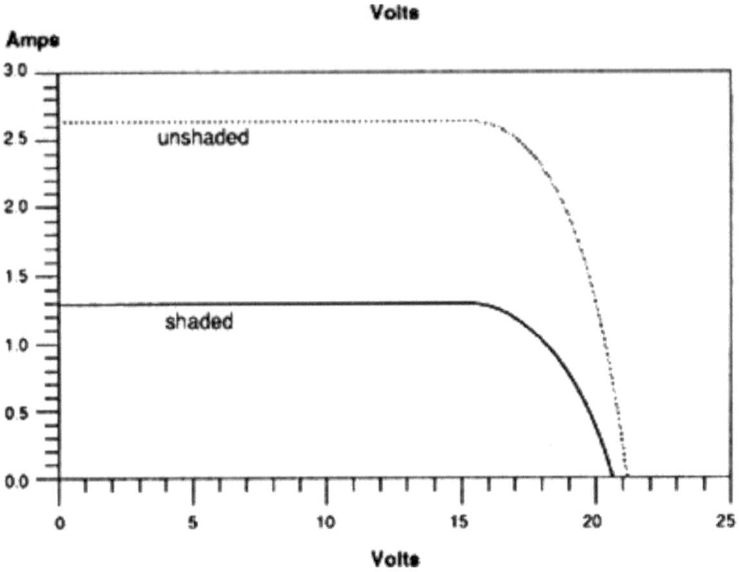

Figure 20.
V-I curve for an unshaded and shaded cell.

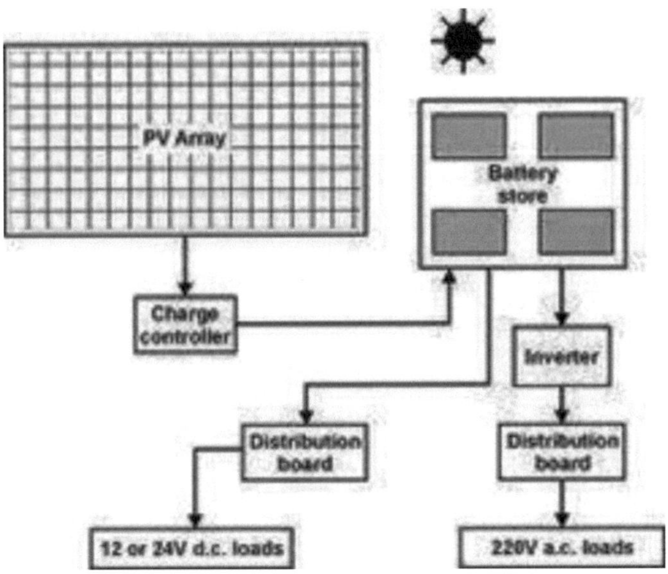

Figure 21.
Standalone system.

electrical vitality provided by a PV board or exhibit. Independent PV frameworks are perfect for remote provincial regions and applications where other power sources are either unreasonable or are inaccessible to give capacity to lighting, apparatus, and different employments. In these cases, it is more savvy to introduce a solitary independent PV framework than to pay the expenses to the neighborhood power organization.

4.8.2 Grid interactive solar PV power system

A grid-connected system is connected to a large independent grid (typically the public electricity grid) and feeds power either directly into a residential or commercial building or back into the grid as shown in **Figure 22**. Grid-associated frameworks differ in size from private (2–10 kWp) to sunlight-based power stations (up to 10 s of MWp). This is a type of decentralized power age [9]. On account of private or building-mounted network-associated PV frameworks, the power created is bolstered into the structure, and any extra power required past what is being created by the framework is conveyed by the matrix. On the off chance that control is delivered in overabundance of what is required by the structure, it is encouraged again into the matrix. The sustaining of power into the network requires the change of DC into AC by an exceptional, framework-controlled sun-based inverter.

4.8.3 Hybrid solar systems

A hybrid sun-powered PV framework consolidates the absolute best attributes of an off-network and a lattice tied framework. These frameworks are associated with the neighborhood utility matrix, yet in addition have a battery back-up framework as shown in **Figure 23**. The battery enables the property holder to store vitality for use amid nondelivering hours (during the evening or amid power outages). These frameworks are perfect for homes where the vitality lattice is questionable on account of harsh climate, a temperamental utility age framework, etc. [8]. This is of specific significance for the individuals who depend on a consistent wellspring of vitality for their home or business.

4.9 We prefer standalone photovoltaic system

Low quality of framework supply (low voltage, fluctuating recurrence, and successive intrusions), high taxes (a lot higher than real expense of supply), uncalled burdens (top hour limitations and impromptu burden shedding), and inert demeanor of State Electricity Boards have constrained numerous ventures to

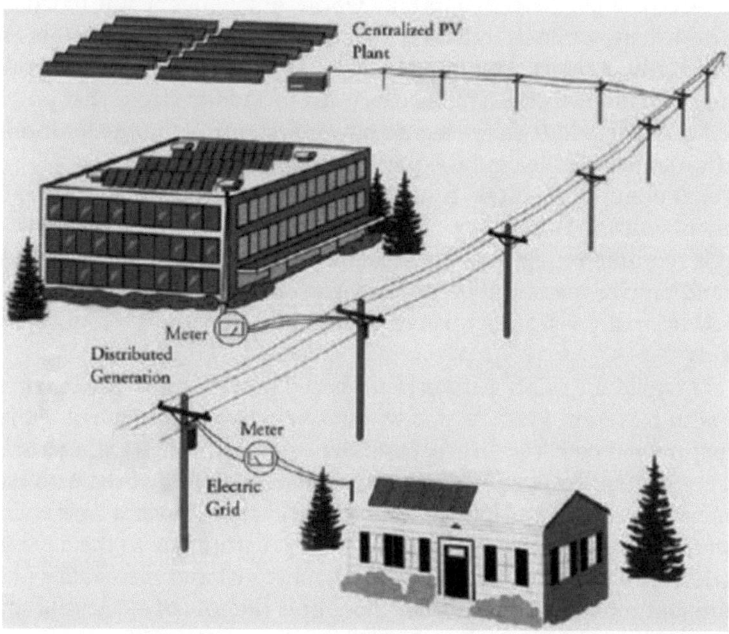

Figure 22.
Grid-connected system.

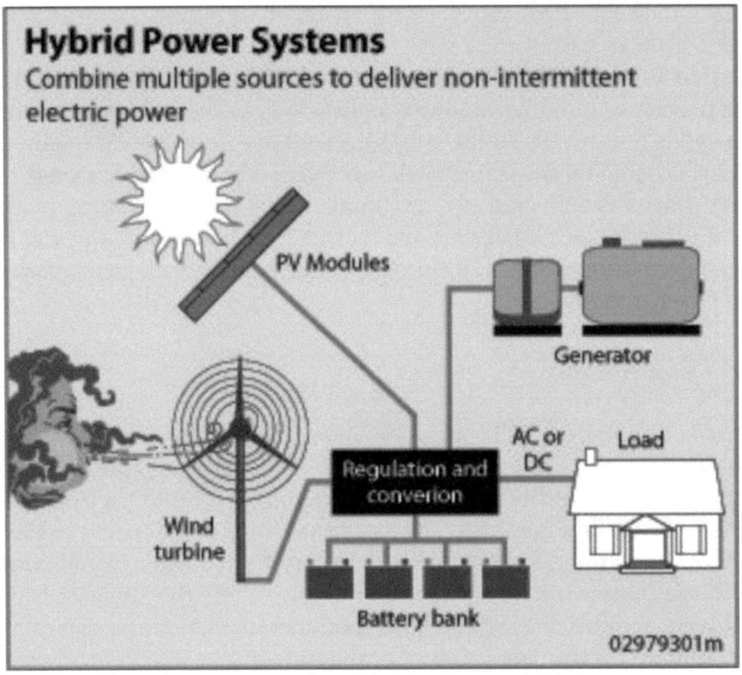

Figure 23.
Hybrid PV system.

segregate themselves absolutely from the state network and be without anyone else. For a solid task of the business, they fundamentally need to utilize hostage age with an excess.

Independent PV frameworks are intended to work autonomous of the electric utility lattice, and are commonly structured and measured to supply certain DC as well as AC electrical burdens. Around the world, independent sun-based establishments are prevalent, while in India, practically all hostage control plants are of the matrix tie. Usually, a smart thought is to begin with little and basic independent sun-oriented PV framework first and afterward to advance from that point. The main benefits of standalone solar electric power systems in remote locations are that they are affordable, reliable, and flexible and their reduced cost [9].

An independent photovoltaic framework does not have the network for reinforcement control. A point by point examination of electric burdens and utilization is required to legitimately structure a framework that will meet the everyday and regular loads and be inside the proprietor/administrators spending plan. It is additionally vital that structures and gear be vitality productive to limit control loads.

There are two kinds of off-network sun-based power frameworks: without batteries and with batteries. Frameworks without battery reinforcement are less difficult to structure and cost less. In any case, they give control just when the sun is sparkling, not during the evening or in awful climate. Frameworks with battery reinforcement give control as long as the battery charge is over a base charge level. These frameworks are progressively intricate to structure due to the day by day and regular variety of illumination (sun-oriented radiation) and the vitality utilization profile (sum and time of day). Another thought is the sort of electrical burden. Direct right (DC) machines can be run specifically from the DC yield from the sun-oriented modules or batteries. Photovoltaic frameworks for rotating current (AC) apparatus utilize an inverter to produce the substituting current. A few frameworks can control both DC and AC apparatus.

Independent frameworks are appropriate as reinforcement or uninterruptible power supplies, particularly for DC hardware or in little networks where matrix control is untrustworthy. Convenient and versatile power frameworks are likewise accessible for remote power circumstances, for example, quick sending, brief burdens, or area evolving loads, for instance, street fix, crisis circumstance, fiasco, salvage, or one-time occasion. Independent sun-based electric frameworks can be planned with a power limit running from 50 Watts to more than 100 KW. The real yield relies upon the insolation (sun-based radiation vitality got on a surface region in a given time) amid the day.

4.9.1 Operational concepts of standalone system

The standalone power system makes use of the solar PV to produce electricity. The configuration of the system is analyzed for various photovoltaic array sizes to operate in tandem with the battery system. The power controller unit will determine the AC conversion of the DC power in relation following the load profile [10, 11]. The charge controller will charge the batteries with energy from solar modules. The main objective of the system is to reduce the cost of operation and maintenance cost by minimizing fuel consumption. A schematic of normal daily operation of a typical solar system can be shown in a series of diagrams in figures below.

4.9.1.1 During day time

In **Figure 24**, solar is the first choice and only source of energy. The inverter converts DC power from the solar PV to AC power for the load. The extra power produced is stored in the battery system.

4.9.1.2 Throughout night

In **Figure 25**, battery is the main wellspring of vitality and sunlight-based PV is off. The inverter changes over DC control from the battery to AC control for the heap. The battery will supply the heap to its most extreme release level.

4.9.2 Elements included in a system of photovoltaic conversion

The main elements that can be included in a system of photovoltaic conversion are the following.

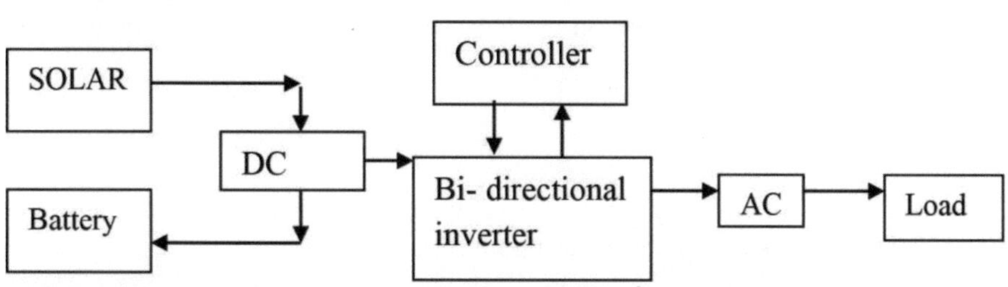

Figure 24.
PV system operation during day time.

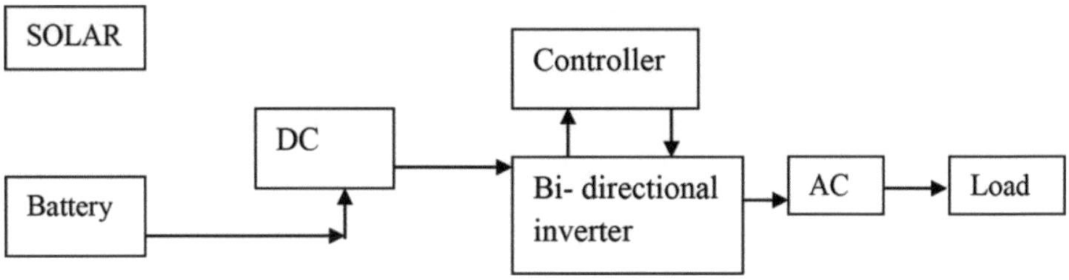

Figure 25.
PV system operation during night time.

4.9.2.1 Batteries

Batteries are a vital component in any independent PV framework; however, they can be discretionary relying on the structure. Batteries are used to store the solar-produced electricity for night time or emergency use during the day. Contingent on the sun-based exhibit setup, battery banks can be of 12, 24, or 48 V and a huge number of amperes altogether. An average battery bank comprises of at least one profound cycle type batteries. The profound cycle batteries are intended to be released and after that energized hundreds or thousands of times. These batteries are appraised in amp hours (AH), which allude to the measure of current that can be provided by the batteries over a particular time of hours. Like sun-oriented boards, batteries can be wired in arrangement and/or parallel to build voltage to the ideal dimension and increment amp hours [11, 12]. The span of the battery bank relies upon the capacity limit required, most extreme release rate, greatest charge rate, and least temperature at which the batteries will be utilized. Batteries can quickly supply huge floods of put away power as expected to begin or run apparatus that the sun powered boards alone could not control.

4.9.2.2 Solar panel

The solar panel is the power source of all photovoltaic installations. It is the result of a set of photovoltaic cells in series and parallel [12]. Solar panel gives power to battery or inverter through charge controller (regulator). The performance of solar PV systems varies with weather conditions due to reduced sunlight exposure on the solar panels, which consequently reduces their output power.

4.9.2.3 Regulator

It is the component to secure the battery against gambling circumstances such as over-burdens and over-releases [13]. The hypothetical definition of the model can be basic, in spite of the fact that it is important to think about the impossibilities to miss discontinuities of the model and the bury the execution with whatever is left of the dissected models.

4.9.2.4 Transformer

A transformer can boost up the ac output voltage from the inverter when needed. Otherwise transformer-less design is also acceptable.

4.9.2.5 Inverter

The inverter allows transforming the DC current to AC. A photovoltaic installation that incorporates an inverter can belong to two different situations, based on the characteristics of the alternating network. An inverter is a gadget that changes DC control from sun-oriented boards or put away in batteries to the standard 120/240 Vac power. Most inverters produce 120 Vac, yet can be outfitted with a stage-up transformer to create 240 Vac. Sun-oriented PV frameworks create direct current, which is put away in batteries. The inverter switches the immediate current forward and backward to create rotating current that is changed into a worthy yield waveform. Inverters come in two fundamental yield plans: sine wave and altered sine wave [14, 15]. The inverter is a noteworthy electronic segment of sun-based PV frameworks and come in evaluations of 50–5500 Watts. It screens control sources and auto chooses among sunlight-based and battery control contingent upon what is accessible.

4.9.2.6 Charge controller

A charge controller anticipates battery over charge and outgassing and is required for charging batteries. A charge controller screens the battery's condition off-charge to protect that when the battery needs charge, the best possible measure of current is given. Interfacing a sunlight-based board to a battery without a charge controller truly chances harming the batteries and makes a potential wellbeing danger. Charge controllers are appraised dependent on the measure of amperage they can process from a sun-oriented board. If a charger controller is rated at 20 amps (A), it means that you can connect a solar panel output of 20 a to the charge controller [16, 17]. A new feature of charge controllers is Maximum Power Point Tracking (MPPT). This is an electronic circuit that improves the effectiveness of sun-oriented boards by augmenting yield control. It enables the charge controller to screen the board's yield and analyzes it to battery bank voltage. At that point, the charge controller changes over the board voltage to the most extreme current for better battery charging. An accuse controller of MPPT improves sun-oriented PV framework execution by around 10%.

4.9.2.7 Metering

A sun-based PV framework meter is like a vehicle measure and is essential for evaluating activity of sun-based PV frameworks. They affirm the battery charging process, show control utilization, battery hold limit, and give chronicled battery information [17]. A meter is normally situated at an advantageous spot in the home. A decent battery meter is a valuable indicative and client administration device.

4.9.2.8 Converter

The situating of a converter between the boards and the batteries will improve the entire photovoltaic establishment, permitting diverse controls from the framework [18]. Contingent upon the connected guideline, the boards will add to the most extreme vitality given to the framework or the ideal vitality for their activity, guaranteeing a productive charge of the battery.

4.9.2.9 Disconnect box

Interfacing sun-oriented PV frameworks to the house and matrix require a distinction box with circuits. Interconnection prerequisites incorporate a utility available box with an obvious primary breaker separate switch, melded between the batteries and other power framework segments to anticipate fires, secure individuals, and gear harm in case of a glitch [18]. The battery banks and sun-powered PV framework additionally need a shared view fixing to the house ground. Having contrasts in ground potential between the house, the batteries and the framework are risky, forcing genuine electrical dangers.

- **Switches and fuses**: these allow the PV system to be protected from accidental shorting of wires and allow power from the PV modules and system to be turned "off" when not required, saving energy and improving battery life.

- **Wiring**: the final component required in PV solar system is the electrical wiring. The cables need to be correctly rated for the voltage and power requirements.

- **Load**: it is the component responsible to absorb this energy and transform it into work.

4.10 Solar radiations

Solar radiation originates at the sun but is measured at the Earth's surface. Before solar radiations had reached the Earth's surface, it must pass through the atmosphere, where it is absorbed, reflected, refracted, and otherwise changed. Together, these components are called the total or global radiation. The sun is a powerful nuclear fusion reactor producing staggering amounts of energy, which is unfortunately dispersed in space and practically all of it is lost. The Earth is 149,596,000 km from the sun, and at this distance, solar flux is relatively small. The energy intercepted by the Earth over 1 year is equal to the energy emitted by the sun in just 14 ms. Energy is radiated by the sun as electromagnetic waves of which 99% have wavelengths in the range of 0.2–0.4 μm [19]. Solar power realization on the crest of the Earth's environment consists of about 8% ultraviolet radiation (short wavelength, less than 0.39 μm), 46% visible light (0.39–0.78 μm), and 46% infrared radiation (long wavelength more than 0.78 μm) as shown in **Figure 26**.

The sun is a big ball of extremely scorching gases, the heat being generated by a variety of fusion reactions. Its diameter is 1.39 × 106 km as that of the Earth is 1.27 × 104 km. The average distance between the two is 1.50 × 108 km. Though the sun is big, it subtends an angle of merely 32 min at the Earth's exterior. This is because it is also at a very large distance [21]. Thus, the beam radiation received from the sun on the Earth is almost parallel. The brightness of the sun varies from

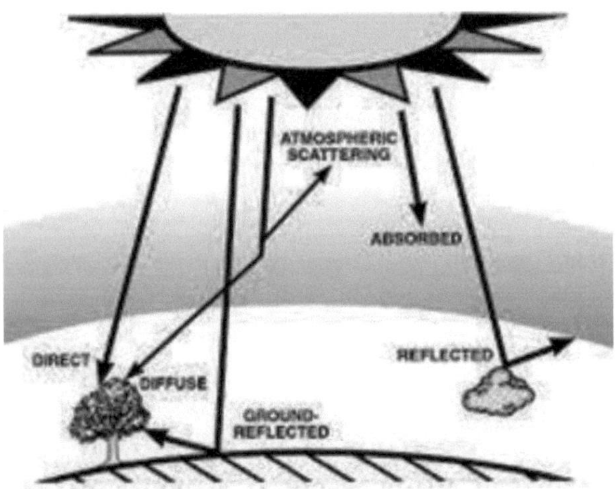

Figure 26.
The distribution of solar radiation.

its center to its edge. The quantity of solar energy falling per instant on unit area, held 90° to the solar radiation outside the Earth's atmosphere when the Earth is at the average distance from the sun is known as the "solar constant." As per the latest measurements, the solar constant has a value of 1.36 KW/m^2 or 1.95 calories/cm^2 per minute.

4.10.1 Earth-sun relationship and insolation

Earth's seasons are constrained by changes in the length and the power of sun-powered radiation or insolation. Both of these variables are thus administered by the yearly change in the situation of the world's hub with respect to the sun. Yearly changes in the situation of the world's hub cause the area of the sun to meander 47° over our skies. Changes in the area of the sun directly affect the power of sun-based radiation. The power of sun-based radiation is to a great extent a component of the edge of occurrence, the edge at which the sun's beams strike the world's surface [22]. In the event that the sun is situated straightforwardly overhead or 90° from the skyline, the approaching insolation strikes the outside of the Earth at right edges and is generally serious. In the event that the sun is 45° over the skyline, the approaching insolation strikes the world's surface at an edge as shown in **Figure 27**. This causes the beams to be spread out over a bigger surface zone of frequency from 90 to 45°. As illustrated, the lower sun angle (45°) causes the radiations to be received over a much larger surface area. This surface area is approximately 40% greater than the area covered by an angle of 90°. The lower angle also reduces the intensity of the incoming rays by 30%.

4.11 Measurement of solar radiations

4.11.1 Universal solar irradiance: pyranometers

The essential instrument used to quantify worldwide sun-powered irradiance is the pyranometer, which estimates the sun's vitality originating from all headings in the half of the globe over the plane of the instrument. The estimation is of the aggregate of the direct and the diffuse sun-oriented irradiance and is known as the worldwide sun-powered irradiance. The most well-known pyranometer configuration utilizes a thermopile (different thermocouples associated in arrangement) connected to a slight darkened retaining surface protected from convective misfortune and protected against conductive misfortunes as shown in **Figure 28**. At the point when put in the sun, the surface accomplishes a temperature corresponding to the measure of brilliant vitality falling on it. The temperature is estimated and changed over through exact adjustment into readout of the worldwide sun-based irradiance falling on the engrossing surface [21, 23, 24]. A legitimately structured

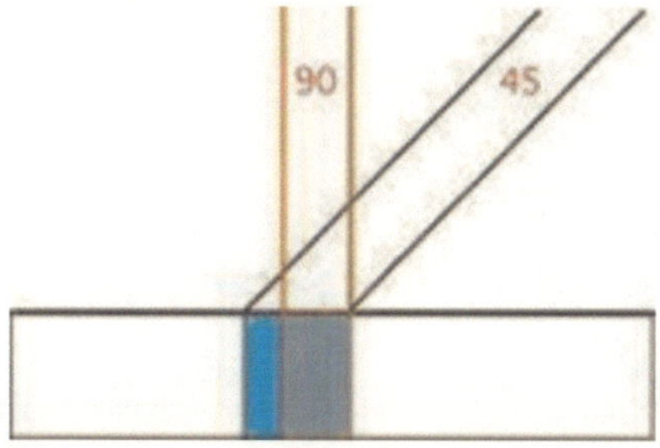

Figure 27.
Effect of angle on the area that intercepts an incoming beam of radiation.

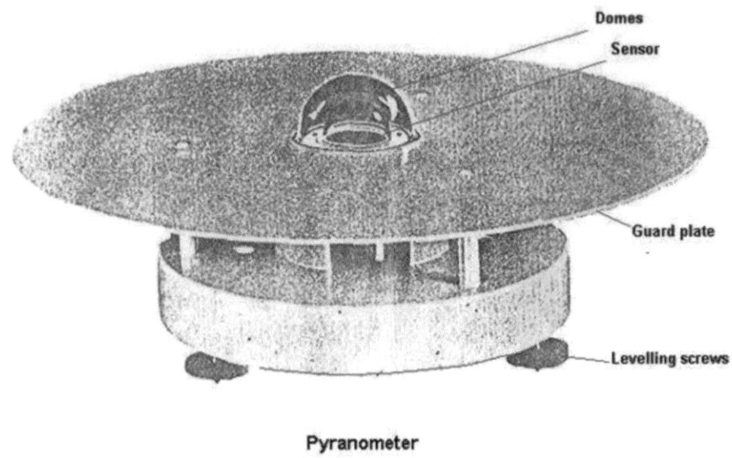

Figure 28.
Pyranometer diagram.

instrument estimates radiation in all the sun-powered wavelengths and its reaction to radiation ought to be relative to the cosine of the edge between the sun and a line typical to the pyranometer safeguard surface.

The common utilization of a pyranometer is for estimation of the worldwide flat sun-oriented irradiance. For this reason, it is set in a level introduction and adequately high over the environment so it has an unmistakable, hemispheric perspective on the whole sky with no shading or reflecting trees or structures inside this field of view [24].

To gauge the immediate typical part of the sunlight-based irradiance, just an instrument called an ordinary rate pyrheliometer (NIP) is utilized. This gadget, shown in **Figure 29**, is basically a thermopile pyranometer put toward the finish of a long cylinder which is gone for the sun [21]. The viewpoint proportion of the cylinder is generally intended to acknowledge radiation from a cone of around 5 degrees. A two-pivot following system is consolidated to keep up the sun's circle inside the acknowledgment cone of the instrument.

Notwithstanding the pyranometer and the ordinary occurrence pyrheliometer, which measures the worldwide and direct sun-oriented irradiance separately, there

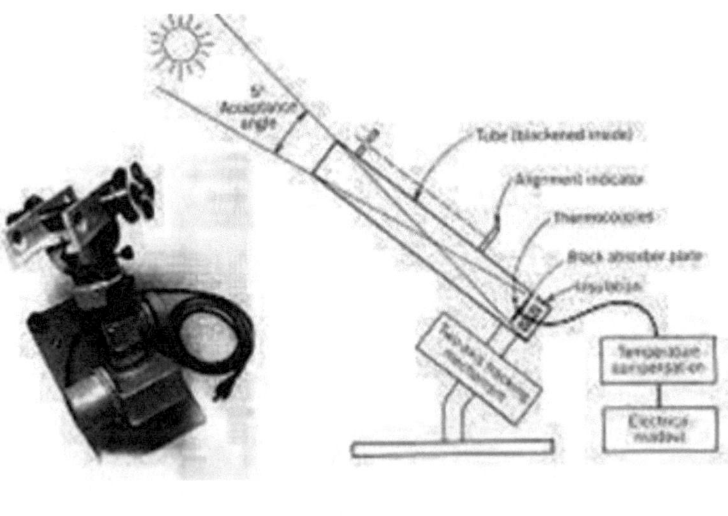

Figure 29.
Pyrheliometer.

is a customary estimation frequently—detailed in meteorological perceptions. This is the "length of daylight." The conventional standard instrument used to quantify this parameter is the Campbell-Stokes daylight recorder. This instrument comprises a glass circle that centers the direct sun-based radiation and consumes a follow on an extraordinary pasteboard card [25, 26]. These recorders have been supplanted in many establishments by photograph indicator enacted "daylight switches".

4.12 Design of a photovoltaic electric power system

Regular strategies for bringing power to the heap focus through long separation voltage transmission/dissemination framework are not in every case financially practical, if the heap focus is situated far from the focal transmission network. In this circumstance, direct transformation and capacity of sun-oriented vitality to power using sun-based cell boards and the batteries offer an option and alluring strategy for giving electrical vitality to such loads. Aside from the sunlight-based boards, the photovoltaic electric power system (PEPS) is made out of various different parts, for example, exhibit structure, mounting outlines, control circuits, wiring and interconnections, stockpiling batteries, and extras. Expenses are likewise acquired for the framework estimating and structure establishment and checkout and testing and upkeep, and so forth, all these can be lumped together as balance of system (BOS). The plan of PEPS for any heap focus needs watchful thought of both sun-powered cell boards and BOS segments [26, 27]. Now we design PEPS for necessary load in Chandigarh University, Gharuan, Mohali.

4.12.1 Electrical energy needs and load pattern

The electrical energy needs of the Chandigarh University (CU), Gharuan for which we have to design PEPS are given below. The load is calculated for Boy's Hostel-2 as shown in **Table 4**.

Calculation of load:

Light points = 5607 × 60 = 336.42 KW

Fan points = 5328 × 80 = 426.24 KW

Plug (5 Amp) = (6921/3) × 60 = 138.42 KW

Plug (15 Amp) = (940/3) × 1000 = 313.33 KW

AC's = 82 × 2500 = 205.00 KW

Motive load = 94.815 KW.

Total campus load = 336.42 + 426.24 + 138.42 + 313.33 + 205 + 94.815 = 1514.225 KW

We consider just the fundamental heap of Boys' Hostel-2 for count for summer season.

The thought of day time load is from 8 am to 4:30 pm while the thought of night load is from 5 pm to 7 pm. To plan a PEPS for the above burdens, most importantly, we build up a strategy to break down everyday electrical vitality needs and burden example of the Boys' Hostel-2 at C.U. Gharuan. The all out size of sun-based cell boards and capacity batteries is resolved from the vitality balance contemplations.

4.12.2 Energy balance considerations

So as to choose the correct size sun-oriented cell boards and capacity batteries for meeting the day and evening time loads, we characterize a vitality balance condition where the absolute vitality accessible from sun powered exhibit (ESA) is adequate to energize the battery (EB) and vitality required by the framework

Sr. No	Name of building	Light points	Fan points	Plug 5 Amp	Plug 15 Amp	AC's
1	Main building	1144	1415	1400	204	30
2	HMCT	400	460	460	100	15
3	Polytechnic	402	468	455	70	10
4	Engineering Block	398	442	450	100	15
5	CBS	405	504	465	112	12
6	Boys' Hostel-1	1413	536	1593	22	Nil
7	Boys' Hostel-2	595	483	1148	Nil	nil
8	Girls' Hostel	700	805	750	264	Nil
9	Workshop	150	215	200	68	Nil
Total		5607	5328	6921	940	82

Table 4.
Total load of C.U. campus.

electrical burden including framework misfortunes (EL), for example, without any reinforcement control supply.

$$ESA = Eb + EL \quad (2)$$

Assuming that the night time load is solely provided by the storage batteries with an overall efficiency factor K1:

$$EB = EN/K1 \text{ where } K1 = \eta D.FU.\eta R.\eta L.\eta B \quad (3)$$

ηD being the solar array diode efficiency, FU the solar array utilization factor, ηR the regulator efficiency, ηL the line loss factor, and ηB the battery Whr efficiency.

Similarly assuming that most of the day time load is directly fed by the solar array except when PD > PSA, the solar array output is given as

$$EL = ED.\alpha/K2 + (1 - \alpha)\ ED/K1 \quad (4)$$

where α is the fraction of day time when PD < PSA and K2 = K1 /ηB is the overall efficiency factor by which the solar array directly drives the load.

The solar array size is expressed in terms of its peak output PSAMP which is obtained from:

$$ESA = PSAMP\sum(f(t)\ dt)I; \text{where I is from } i = 1 \text{ to P} \quad (5)$$

where f(t) = ESA/PSAMP over the ith segment (dt)I during the day time load. z is the number of time segments in the day.

The capacity (CB) in Ahr of the storage battery at the rated load is determined by the daily night load and a part of day time load for which sufficient storage is to be provided as protection against cloudy weather, thus leading to

$$CB = xPN + yPD/[\eta LVD\ (CD/100)] \quad (6)$$

where x and y are the periods for which storage are to be provided for different loads, VD is the average voltage of discharge of batteries, and CD is the maximum permissible depth of discharge in percentage.

4.12.3 Balance of system components

As referenced before, separated from the sun-based cell boards, the PEPS includes various parities of framework parts, for example, (a) cluster module mounting outlines, outline backings, and establishments; (b) electrical control circuits, load executives and power molding hardware, wiring interconnections, and so on; and (c) stockpiling batteries, racks, and venting types of gear, and so on [22].

Further expenses are likewise caused for the establishment and checkout of module test and investigation, framework estimation, and structure bundling upkeep, and so on; the whole of all these expenses, lumped as BOS cost, should be considered for plan of PEPS. An important factor contributing to the highly nonlinear behavior of BOS cost /peak W with the PEPS size is the nature of variation of the storage batteries cost/Ahr with the capacity in Ahr of battery.

4.13 Design requirements

A module can be made by attaching one cell to another with metal shaped from solid wire or solid ribbons. The interconnections can be rigid or flexible to contend with movement within the array produced by thermal expansions and other forces. All connections should provide for the lowest possible resistance and least possible interference with PV performance. Thus designers attempt to keep such connection short, reducing cross sectional area against increasing resistance. The output from an array is tied to a collecting conductor called a bus.

4.13.1 Placing of cells

Placements of cells in the array and cell's shape are important. Overall panel efficiency as measured by voltage per unit area drops as the space between cell increases. Large cells are not often to boost packing efficiency (i.e., the need for a maximum cell to panel area ratio). Cell sizing is an important aspect in creating a module with desired electrical properties. The current available from a cell varies with the size of the cell and voltage remains constant. For large voltage, many small cells should be connected in series [19]. Round cells are used that have been cut in half and then placed into an array in an offset pattern to get more of them into a unit area. This increases the packing density of the cell. It can also be increased by square or hexagonal cells. Cells are placed as closely or as possible and cannot be allowed to touch as they will short out electricity [30]. Extra space must be allowed between the cells to accommodate thermal expansion.

4.14 Array support

There is more to array building that resourcing electrical need individual solar cells are fragile groups of them can be equally fragile. Each module must be able to hold up to the rigour of assembly and disassembly. An array must be able to withstand mild loads, mechanical movements, and stresses induced by temperature changes. Part of a module's support is the form of the transparent cover applied to it [31]. The primary application of the cover is to protect the PV module against conditions with oxygen, humidity, dust, and precipitation.

4.14.1 Size of array

Size of solar cells may be ranging from about 1 mm to over 100 mm in diameter. The thickness range for the most common silicon cells is 0.2–0.4 mm.

We developed a very simple semi-empirical rule for the selection of the size of array.

$$Pph = (LH + LHd/Cr\ Bb \times 100)/X \quad (7)$$

where Pph is the array size in peak watts. X is the annual average equipment peak hours per day that is the annual average watt-hours received per weak hour of flat plate photovoltaic module per day. L is load rating in watts. H is the hours of operation per day. d is the number of days of storage required. Cr is the charge recovery period of the period and Bb is the watt-hour efficiency of the battery.

The value of X is directly dependent on the total insolation received by the panel at the site of installation. The value of X can be estimated as

$$X = EXm/12; \quad \text{With } Xm = \eta ov\ Im/\eta m \quad (8)$$

where ηov is the overall efficiency of the system, Im is the average insolation on a horizontal surface of the location, and ηm is the module efficiency. The overall efficiency is the product of the module efficiency and the balance of system efficiency, including the power conditioning efficiency, temperature coefficients of efficiency, and so on.

4.14.2 Solar panels

Most of the silicon solar cells employed for terrestrial applications are round of 5 cm diameter and a thickness of 0.3–0.5 mm. the tendency is toward large diameters. A cell of 5 cm diameter with a surface area of about 20 square cm delivers in full sun and at room temperature a power of 0.2 W at 0.45 Volts. For higher power or higher voltage, a number of cells must be assembled into a panel [31, 32]. For instance, to double power at constant voltage, two cells are connected in parallel. By connecting a number of cells in parallel and series, it is possible to provide any amount of power at a desired voltage.

4.14.3 Battery storage

The simplest means of storage on a smaller moderate scale is in electric storage battery. Solar cells produce the direct current required for battery charging. The stored energy can then be delivered as electricity to the local load when needed. A battery is a combination of individual cells. A cell is the elemental combination of materials and electrolyte constituting the basic electromechanical energy storer. A battery can also be thought as a block-box into which electrical energy is put, stored as electromechanical energy, and later regained as electrical energy. Primary batteries are nonrechargeable while secondary batteries can be recharged again and again. So secondary batteries are of chief interest for solar electrics. Examples of secondary batteries are lead-acid, nickel-cadmium, iron-air, nickel-hydrogen, zinc-air, sodium-sulfur, sodium-chlorine, etc.

Energy efficiency of a battery is defined as

$$\eta energy = \int I1\ E1\ dt / \int I2\ E2\ dt \quad (9)$$

where I1 = battery discharging current for a period of 0 to t1. I2 = battery charging current for a period of 0 to t2. E1 = Battery discharging terminal voltage. E2 = Battery discharging terminal voltage.

Cycle life is the number of times the battery can be charged and discharged under specified conditions and this may vary greatly with the depth of the discharge. Deep discharge tends to result in short cycle life.

4.15 Design calculations

Site information:
Proposal site for the solar plant is Chandigarh University, Gharuan, Mohali, Punjab. Its coordinates are
Latitude: 30°46′5″ N
Longitude: 75°34′36″ E

Now by using the above relations, we can design the PEPS as below.

Type of load	Quantity	Wattage	Total wattage	Hours of operation	Total KWhr
Light points	593	60	35,580	17	656.880
Fan points	483	80	38,640	7	249.060
Total					905.940

- Total roof area = 1050.1416 m^2

During summer, solar energy is available for 8 h per day. Therefore, array size can be calculated using a natural energy loss factor of 0.85 [9].

- Array size = (0.85 × 905.94)/8

= 96.256 KW

= 97 KW (approximately)

- We use 150 W panel of 1.5 m^2. Therefore, the number of solar panels (Ns) required is calculated as

Ns = total load in watts/rating of single panel in watts

= 97 x 103/150

= 646.67

Ns = 647 (approximately).

(Based on the roof area we need: 1050.1416 m^2/1.5 m^2 = 700. Therefore, we have enough space for the accommodation of 647 panels.)

- We use 12 Volts, 17 Ahr lead acid battery. The capacity of storage batteries CB in Ahr is determined as

CB = total KWhr/voltage of single battery

= 905.94 KWhr/12 Volts

= 75.495 KAhr

- The number NB of lead acid batteries required can be calculated as

NB = CB/Ahr rating of single battery.

= 75.495 KAhr/17 Ahr

= 4440.88

= 4441 (approximately)

- We use 12 V, 20 Ampere charge controller. The rating of the charge controller is in amperes and can be calculated as

= Total load in W/12 Volt

= 97 × 103 W/12 Volts

= 8.0833 × 103 Amp

= 8083.333 Amp

= 8083 Amp

- Number of 12 V, 20 Amp charge controllers (NC) required can be calculated as

NC = 8083/20

= 404.15

= 404 (approx.)

For 97 KW load we need 97 KW inverter.

Now the total BOS cost can be calculated as

Cost per watt is $0.70 or Rs 37.667 [14].

Cost of solar panels C = total load in watts × cost per watt

= 97 × 103 W × Rs 37.667

C = Rs 3,653,699

Cost of battery can be calculated as:

= NB × cost of one battery

= 4441 × $34

= 4441 × 1982.54

= Rs 8804460.14

Cost of charge controller can be calculated as

= NC × cost of one charge controller

= 440 x Rs 1899

= Rs 835,560

Cost of 97 KW inverter can be calculated as

= total load in KW × cost per KW

= 97 × $2000

= 97 × Rs 107,620

= Rs 10,439,140.

Total cost = cost of solar panel + cost of battery + cost of charge controller + cost of inverter.

= 3,653,699 + 8804460.14 + 835,560 + 10,439,140

= Rs 23732859.14

To take into account the cost of wiring, junction box, etc., 20% of the total cost is added to get the total cost of the project.

= 20% Rs 23732859.14

= Rs 4746571.828

Therefore, the total cost of the project C = Rs 23732859.14 + 4746571.828

C = Rs 28479430.83

If we purchase the energy from the utility, we have to pay

= total demand in KWhr × price of one unit

= 905.94 KWhr × Rs 7/unit

= Rs 6341.58

Total cost per year is

D = Rs 6341.58 × 365

D = Rs 2314676.7/year

Payback period

It is the period of time required to recoup the expense of a venture. The recompense time of a given speculation or venture is a critical determinant of whether to embrace the position or undertaking, as longer compensation periods are regularly not attractive for speculation positions.

Payback period = cost of project/annual cash inflows

The payback period can be calculated using the following equation:

C – ND = 0

Or N = C/D

where C = Rs 28479430.83

D = Rs 2314676.7/year

Therefore,

N = Rs 28479430.83/Rs 2314676.7

N = 12.3

Therefore, the cost of the project installation can be paid back or recovered in 12 or 13 months.

5. Results

From the structure system, roof top methodology has accommodated the establishment of the sun powered boards at Boys' Hostel-2 C.U. This methodology is the coordination of the boards to the top of the structure. This methodology is given as it replaces the regular rooftop while enabling the normal daylight to channel through. As a rooftop, it servers as auxiliary and climate condition prerequisites by giving basic quality and solidness; it secures against harms like substance and mechanical harm, averting against flames, and ensuring against downpour, sun, wind, and dampness; it permits heat assimilation and warmth stockpiling; it controls the dissemination of light and so on; notwithstanding these highlights, it fills in as a power generator through gathering some portion of the electrical burden necessities of the structure.

In view of the very particular nature of both sunlight-based cells and capacity cells, indicated burdens can be controlled independently by individual rooftop-top PEPS for meeting similar vitality needs of the diverse loads as referenced previously.

6. Conclusion and future scope of the work

It is normal that with present increasing speed in the endeavors with respect to makers, creators, organizers, and utilities with satisfactory governmental support, PV frameworks will within the following two decades involve a position of pride in the nation's capacity part, guaranteeing ideal usage of the vitality specifically from the sun around the year. Plainly, the SPV framework can give some help toward future vitality requests. This PV framework comprises PV exhibit with vitality putting away gadgets and power electronic gadgets that have been talked about in this undertaking work to accomplish an effective and cost focused framework design so that sun-based power sources could improve the life of individuals particularly in provincial regions where power from primary lattice has not come yet [33]. The procedure received appears to be tasteful for deciding the conceivable required vitality from the sun-based board for a self-assertively picked zone. The complete BOS cost determined appears to be palatable for the proposed structure. The surplus vitality produced can be utilized when sun is not accessible.

6.1 Future scope of the work

Later on, we will figure out the quantity of PV exhibits and cost of the framework which can satisfy the heap need of all grounds. In the beginning, we have not considered the climate control system load. In future, we will incorporate the heap of forced air systems. A point by point cost investigation can be made considering carbon credit to demonstrate whether it is monetarily feasible or not since the execution of PV framework is firmly reliant on misfortune factors, for example, shading, PCS misfortunes, bungle, PV exhibit temperature rise, and so on. There is a need for checking on these misfortune elements to assess and investigate precisely the execution of PV framework. This framework can be structured with additionally some other electrical machines like channel for stifling the swells. A point by point execution examination of the present framework can be done to demonstrate its unwavering quality as a future study. Sun-based PV is an innovation that offers an answer for various issues related with nonrenewable energy sources. It is spotless decentralized, indigenous, and does not require consistent import of an asset. What's more, India has among the most noteworthy sun-oriented irradiance on the planet which makes sun-based PV even more appealing for India. The territories of Orissa and Andhra Pradesh likewise house probably the best quality stores of silica. India has a substantial number of cell and module producers. Despite every single above preferred standpoint, Indian Photovoltaic Program is still in the earliest stages. One reason could be nonappearance of basic, activity arranged, and forceful PV approach of the nation both in state and focal dimension. All the more rapidly we do it with the experts, the more we ensure our future vitality security.

Acknowledgements

I am thankful to Chandigarh University management for giving me an opportunity to undergo my project in this esteemed institute. My most thanks to Dr. Inderpreet Kaur, Professor and Head, Department of Electrical Engineering, Chandigarh University, Gharuan, for continuous support and motivation.

A. APPENDIX

Specification of solar panel used

EPCOM polycrystalline solar panel

Wattage of single solar panel = 150 Watt

Cells per module = 36

Maximum power voltage = 18.28 V

Maximum power current = 8.21 Amp

Open circuit voltage = 21.9 V

Short circuit current ISC = 8.93 Amp

Cell efficiency = 17%

Module efficiency = 14.9%

Cell size (mm) =156 × 156

Dimensions = 1480 × 680 × 35

Weight = 11.6 kg

Price of single panel = $ 0.70

Specification of sealed lead acid battery

Nominal voltage = 12 V

Nominal capacity = 17 Ahr, 204 Whr

Maximum charging current = 5.1 Amp

Maximum discharging current = 255 Amp

Dimension = 181 mm × 76 mm × 167 mm

Weight = 6150 g

Energy density = 3 Whr/Kg

Price = $34.00

Others

Maximum number of panels in series = V inverter/V panel

Maximum number of panels in parallel = I inverter/I panel

Author details

Harpreet Kaur and Inderpreet Kaur*
Electrical Engineering Department, Chandigarh University, Gharuan, Mohali, India

*Address all correspondence to: hod.eee@cumail.in

References

[1] Bhadai L. Study in techno economic aspects of power generation from agri wastes in India [PhD thesis]. Patiala: Mechanical Engineering Department, TIET; 2005

[2] Rai GD. Non-Conventional Energy Resources. New Delhi: Khanna Publishers; 2004

[3] Janssen P, Myrzik J, Kling W, Reinders L. Technical feasibility study for a solar energy system at Amsterdam Airport Schiphol (AAS). In: International Conference on Renewable Energies and Power Quality (ICREPQ'10), Granada, Spain; March 23–25, 2010. pp. 909-914

[4] https://www.eia.gov/outlooks/ieo/. International Energy Outlook (IEA) accessed on 8th January, 2019

[5] Chen Q, Gua Y, Tang Z, Sun Y. Comparative environmental and economic performance of solar energy integrated methanol production systems in China. Energy Conversion and Management. 2019;**187**:63-75

[6] Augenbraun JJ. Energy From the Sun: A Solar Feasibility Study for Macquarie University. Available from: http://digitalcollections.sit.edu/isp_collection/868

[7] Ahmed A, Hafez A, Alblawi A. A feasibility study of PV installation: Case study at Shaqra University. In: The 9th International Renewable Energy Congress (IREC 2018). IEEE; 2018. 978-1-5386-0998-9/18/$31.00 ©2018

[8] Grunau B, Greg Egan PE. Solar Energy Feasibility Study. Available at: http://www.cchrc.org/sites/default/files/docs/CCHRC_Solar_Feasibilty_Study.pdf

[9] Joshi KA, Pindoriya NM. Impact investigation of rooftop solar PV system: A case study in India. In: 2012 3rd IEEE PES Innovative Smart Grid Technologies Europe (ISGT Europe); Berlin; 2012. 978-1-4673-2597-4/12/$31.00 ©2012

[10] Kowli A, Raj RP, Bandam A. Assessing the feasibility of large-scale adoption of solar power in the residential sector. IEEE; 2016. 978-1-4799-5141-3/14/$31.00 c 2016

[11] Lee M. Economic feasibility analysis and policy implication for photovoltaic system at cohousing in Korea. Renewable Energy. https://doi.org/10.1016/j.renene.2018.11.109

[12] Pandey A, Agrawal HP. Design of hybrid power system for an academic institution. International Conference on Applied and Theoretical Computing and Communication Technology. IEEE; 2015. 978-1-4673-9223-5/15/$31.00_c 2015

[13] Putri RK, Wardhany AK, Astuti YDRW, Hudaya C. Building integrated photovoltaic for rooftop and facade application in Indonesia. International Conference on Green Energy and Applications. IEEE; 2018. 978-1-5386-5236-7/18/$31.00 ©2018

[14] Raviprasad V, Ravindra KS. Feasibility study of a small SPV power plant added to existing rooftop BTS. In: 2012 IEEE International Conference on Power Electronics, Drives and Energy Systems. December 16–19, 2012; Bengaluru, India; 2012

[15] Kaur H, Kaur I. Energy Return on Investment (EROI) analysis of 2KW Solar Photovoltaic System. Springer Sponsored International conference on Green Technologies For Power Generation, Communication And Instrumentation (ICGPC,2019) on 3-4th April, 2019 ISBN NO:978-93-5254-979-5

[16] Lodin O, Kaur I, Kaur H. Designing an effective and efficient solar tracking

system to overcome the drawbacks of conventional P&O, International conference on Green Technologies For Power Generation, Communication And Instrumentation (ICGPC,2019) on 3-4th April, 2019 ISBN NO:978-93-5254-979-5

[17] Cabeza LF, de Gracia A, Pisello AL. Integration of renewable technologies in historical and heritage buildings: A review. Energy and Buildings. 2018;**177**: 96-111

[18] Mewes D, Monsalve P, Gustafsson I, Hasan B, Palén J, Nakakido R, et al. Evaluation methods for photovoltaic installations on existing buildings at the KTH Campus in Stockholm, Sweden, International Conference—Alternative and Renewable Energy Quest, AREQ 2017, 1–3 February 2017, Spain. Energy Procedia. 2017;**115**:409-422

[19] Wang Q, Zhou Y, Gao S. Feasibility analysis of solar water heating system in rural areas. 10th International Symposium on Heating, Ventilation and Air Conditioning, ISHVAC2017, 19–22 October 2017, Jinan, China. Procedia Engineering. 3852-3859

[20] Ashhab M'd SS, Kaylani H, Abdallah A. PV solar system feasibility study. Energy Conversion and Management. 2013;**65**:777-782

[21] Eldin SAS, Abd-Elhady MS, Kandil HA. Feasibility of solar tracking systems for PV panels in hot and cold regions. Renewable Energy. 2016;**85**:228-233

[22] Okoye CO, Oranekwu-Okoye BC. Economic feasibility of solar PV system for rural electrification in Sub-Sahara Africa. Renewable and Sustainable Energy Reviews. 2018;**82**(3):2537-2547

[23] Raugei M, Fullana-i-Palmer P, Fthenakis V. The energy return on energy investment (EROI) of photovoltaics: Methodology and comparisons with fossil fuel life cycles. Energy Policy. 2012;**45**:576-582

[24] Hall CA, Balogh S, Murphy DJ. What is the minimum EROI that a sustainable society musthave? Energies. 2009;**2**:25-47

[25] Halder PK. Potential and economic feasibility of solar home systems implementation in Bangladesh. Renewable and Sustainable Energy Reviews. 2016;**65**:568-576

[26] Kamali S. Feasibility analysis of stand alone photovoltaic electrification system in a residential building in Cyprus. Renewable and Sustainable Energy Reviews. 2016;**65**:1279-1284

[27] Kamran M et al. Implementation of improved Perturb & Observe MPPT technique with confined search space for standalone photovoltaic system. Journal of King Saud University - Engineering Sciences. 2018. https://doi.org/10.1016/j.jksues.2018.04.006

[28] Mohanty A, Ray PK, Viswavandya M, Mohanty S, Mohanty PP. Experimental analysis of a standalone solar photo voltaic cell for improved power quality. Optik - International Journal for Light and Electron Optics. 2018;**171**:876-885

[29] Shukla AK, Sudhakar K, Baredar P. Design, simulation and economic analysis of standalone roof top solar PV system in India. Solar Energy. 2016;**136**: 437-449

[30] Arya A, Ahmad MW, Anand S. Online monitoring of power extraction efficiency for minimizing payback period of solar PV system. IEEE; 2015. pp. 2863-2868. 978-1-4799-7800-7/15/ 2015

[31] Hayat M, Shahnia F, Arefi A, Iu H, Fernando T. Comparison of the economic benefits and the payback periods of rooftop solar panels in Australia. International Conference on Power Generation Systems and Renewable Energy Topologies. IEEE;

2017. pp. 113-117. 978-1-5090-5353-7/15/2017

[32] Ivanova IY, Tuguzova TF, Khalgaeva NA. Comparative analysis of approaches to consider rationale of use of solar panel plants for power supply of off-grid consumers. In: International Ural Conference on Green Energy. IEEE; 2018. pp. 75-78. 978-1-5386-4936-7/18/2018

[33] Wijesuriya DTP, Wickramathilaka KDSH, Wijesinghe LS, Vithana DM, Ranjit Perera HY. Placing reflectors for reducing payback period of solar PV for smart buildings. International Conference on Industrial Informatics. IEEE; 2018. pp. 480-485. 978-1-5386-0837-7/17/2017

[34] Sharma V, Kaur H. Design and Implementation of Multi Junction PV Cell for MPPT to Improve the Transformation Ratio. International Conference on Green Technologies For Power Generation, Communication And Instrumentation (ICGPC,2019) on 3-4th April, 2019. ISBN NO:978-93-5254-979-5

Chapter 4

Efficiency, Energy Saving, and Rational Use of Energy: Different Terms for Different Policies

Nino Di Franco and Mario Jorizzo

Abstract

In recent years, the increasing interest for energy efficiency has multiplied the number of players and the issuing of legislative documents, so the very notion of "efficiency" has taken different meanings in a more or less wide range of definitions, sometimes overlapping between them. These definitions often evoke different concepts such as "energy saving," "rational use of energy," "efficient use of resources," "reduction of consumptions," etc., in an amalgam shadowed by ambiguous interpretations. This paper proposes a clarification of the different expressions by defining their functional and conceptual boundaries and interrelationships, focusing the attention on the energy aspects, and leaving out other dominions that might govern or accompany that variable, such as, sustainability, competitiveness, economy, etc. The issue is not merely lexical or taxonomic. In fact, the strict definition of a concept defines its area of interest, and the decision-maker, when issuing a measure, should choose from his portfolio of available tools only those consistent with the involved domain.

Keywords: efficiency, energy saving, rational use of energy, definition, classification

1. Introduction

In recent years, the increasing interest for energy efficiency has multiplied the number of players and the issuing of legislative documents, so the very notion of "efficiency" has taken different meanings in a more or less wide range of definitions, sometimes overlapping between them [1, 2]. These definitions often evoke different concepts such as "energy saving," "rational use of energy," "efficient use of resources," "reduction of consumptions," etc., in an amalgam shadowed by ambiguous interpretations. The Energy Efficiency Plan 2011 of the European Commission [3] quotes, for example:

> *Technically, "energy efficiency" [4] means using less energy inputs while maintaining an equivalent level of economic activity or service; 'energy saving' [5] is a broader concept that also includes consumption reduction through behavior change or decreased economic activity. In practice the two are difficult to disentangle and—as in this communication—the terms are often used interchangeably.*

According to IEA [6], "energy efficiency" is a concept that can be difficult to define since it can mean different things to different people. One difference of opinion usually lies in whether energy efficiency encompasses only the technical efficiency of an energy service, i.e., the energy consumed as a result of a technological performance, or whether non-technical factors such as behavior are included in the interpretation of energy efficiency.

Directives 2006/32/EC and 2012/27/EU contain their own definitions of "energy efficiency" and "energy saving," but ambiguity between the terms persists. This paper proposes a clarification of the different expressions by defining their functional and conceptual boundaries and interrelationships, focusing the attention on the energy aspects, and leaving out other dominions that might govern or accompany that variable, such as, sustainability, competitiveness, economy, etc. The issue is not merely lexical or taxonomic. In fact, the strict definition of a concept defines its area of interest, and the decision-maker, when issuing a measure, should choose from his portfolio of available tools only those consistent with the involved domain. There would otherwise be the risk of using resources to promote "efficiency" using "energy-saving" tools possibly insufficient or even incompatible with the desired goal.

The lack of strict definitions in the field of the rational use of energy is, on one side, a source of uncertainty in the identification of targets, in their degree of achievement and selection of most suitable policy tools, and, on the other side, a reason of unwilled policy bias toward specific option and result [7, 8]. The main weaknesses of the energy policies developed in these conditions are the lack of criteria for monitoring performance, the lack of adequate financial assistance, and inappropriate communication in terms of message as well as targeted audience.

An example of great importance that corroborates our position is linked to the effects of the Green Deal (GD) [9], a vast plan to promote energy saving in homes launched in the UK in 2013, and which has shown signs of suffering since the first year of application (e.g., compared to a target of 2 million homes to be retrofitted each year, only 6000 had been retrofitted every year by the end of 2016). The vast literature produced in this regard has identified the following, among the various causes:

1. The GD did not require that the financeable efficiency measures should achieve given levels of energy savings nor provide criteria for monitoring the performances, thereby introducing uncertainties on the degree of achievement of the targets and on the corrective measures to be introduced (an energy saving must be measurable).

2. In a first phase (2013–2014), no state subsidies were envisaged that would make the efficiency improvement measures profitable (the net present value of energy savings must be positive).

3. In a second phase (2014–2015), state funds were made available, but not to the extent required by demand or in a long-term perspective (energy savings must be stable over time).

4. The government team that introduced the GD was composed of generalist officials with no experience in the field of energy efficiency, and above all without experts in the social-psychological, marketing, and communication fields (a given goal should require dedicated professionalism).

5. The GD could finance energy efficiency measures (e.g., insulation of walls), energy waste reduction (e.g., draft proofing), and use of renewable sources (e.g., solar panels) altogether. These actions aim at conceptually different goals and should have required different tools to be implemented.

A proper definition of the terms "energy saving," "efficiency," and "rational use" will therefore make correlations possible like

"area of interest → tool"

as support for the policy-makers when establishing the principles of an action plan and for the analysts to check the results and the inner consistency between the goals and the means used to achieve them.

2. The classification issue

In recent decades many pieces of legislation on the issue of energy conservation at EU and national levels have been produced. Their stated goal is the improvement of the use of energy by end users through both prescriptive standards and direct/indirect financial support. Over time, such legislative tools have limited temperature in homes at 20°C, provided tax deductions for energy-saving interventions in buildings, granted energy efficiency certificates for measures in the industry sector, introduced labeling and energy performance certification, stimulated the market for the ESCOs, stated the appointment of an energy manager, supported the practice of energy audits and energy performance contracts, etc., thus creating a fertile ground for the identification of additional areas—in productive activities and in social life—of improvement. This legislative and regulatory process [10] is still in progress, and Italy and the EU Member States consider it as a cornerstone of their energy policy. On the other hand, energy can be saved even when turning off the lights when leaving a room, recovering heat in a production process, buying four-star appliances, installing a cogeneration plant, joining the local district heating, using stairs instead of elevators or bikes instead of cars, eliminating drafts under doors at home, connecting an inverter to an electric motor, becoming vegetarians, funding ads and documentaries on TV to create awareness among consumers, inserting the chapter "How to Use energy" in the books for elementary schools, and installing smart meters or modifying the wing profile of an aircraft or favoring the recycle of glass. Given the current regulatory framework, it would be difficult to identify, between those listed above, the "rational use of energy" or "energy-saving" or "energy efficiency" measures. The task we want to deal with in the next chapter is the survey of these families but also of other and different ones, in which the various measures listed above are logically contained.

3. Limiting energy consumptions

Any kind of measures able to save energy, e.g., those listed in the previous chapter, could belong to two classes: technical and nontechnical measures. All the initiatives related to plants or machineries with better performance than those previously installed may belong to the first family, while initiatives derived from behaviors of certain social classes (e.g., public employees, workers, students, families) or from the way certain productive processes are managed may belong to the second. The family of the technical measures could be further subdivided by the energy carrier (electricity, steam, fuel, etc.), by the primary source saved (natural gas, oil, coal, biomass, etc.), etc.

Another possibility is to classify the measures according to the kind of approach, whether top-down or bottom-up. Top-down measures "command and control" imposed by a higher authority; bottom-ups are those stemming from a free decision of the final users.

Another criterion might be related to the kind of final uses, for example, buildings (walls or the HVAC plants), industry (the utilities or the process), the tertiary sector, and transports and, again according to the production chains (paper, glass, textile, etc.), by types of services provided (by schools, offices, retail, etc.) or kind of transport means (land, air, naval).

Other methods could address the complexity of the measures (from "no cost good housekeeping" practices such as turning off lights to complex project with relevant financial implication)

The criteria partially exposed above, in the absence of a unifier element, seem, however, biased toward a simple cataloging, able in case to put order in the great family of "measures to limit energy consumptions" but unable to provide added values for the decision-maker/legislator when establishing an integrated policy.

A new method based on a classification by areas of interest is presented below; uniform measures will be referred to any given field, to be implemented with consistent and dedicated resources and tools.

4. Energy saving

Let's start from the "energy-saving" concept through a clear and precise definition interpreting the expectation of a given policy, namely, an available and effective tool for the reduction of the energy consumptions in a framework of increased competitiveness, sustainability, and alleviation of the trade balance with foreign countries.

First of all let's ask ourselves if, reducing the energy consumption in a given context from a value E1 to a value E2, one could call "energy saving" the difference of E1-E2. We think it is possible as far as the following conditions are met.

First, "energy saving" should be voluntary and programmable. As such, it must come from a plan considering the final users' consumption profile, the technological offer, and the trend of energy markets. If a reduction of the demand came from nonvoluntary factors instead—for example, thanks to the favorable climatology or to the market dynamics making low-energy products or services more attractive in certain periods—such reduction might constitute a lucky contingency, but it couldn't be called "energy saving" because the feature of planning, linked to any policy action, would be missing.

Second, "energy saving," meaning the difference between an ex ante and an ex post consumption, must be measurable: the decision-maker, whether public or private, must be able to precisely determine the quantitative effect E1-E2 that the chosen initiative, once realized, will be able to produce in order to assess the adequacy to achieve a given objective, to follow the evolution of the results produced in time and to compare the actual results against the amount of resources fielded. In this regard, let's consider the typical asymmetry between the ex ante and ex post measurements: when implementing an initiative to rationalize a given device, consumptions ex ante are certainly measurable, while the future ones will only be alleged and uncertain, and, if possible, they can only be estimated using engineering formulas. However, when the initiative is put in place at present, consumptions ex post are surely measurable; the effects are then determined (in the absence of a meter already installed on the machine), counting ex ante consumptions through empirical and/or statistical methods. The lack of a meter measuring ex ante or ex post consumptions necessarily causes uncertainty in the calculation of actual savings.

From the need to measure energy savings, one can argue that the technological initiatives, whose ex ante and ex post consumptions are certainly measurable, fall certainly within the definition; it will not be so easy for non-technological initiatives (it can be difficult to measure the energy saving derived form an increased "energy culture" of the end users or produced by an information campaign etc.).

Third, "energy saving" should produce net positive energy savings compared to the ex ante situation. While measurability is a precondition, it is also necessary that the balance between ex ante and ex post consumptions is positive, and therefore it should always be E1-E2 > 0. This condition is apparently tautological (talking about "savings" the difference of E1-E2 should de facto be greater than zero) because the assessment of the energy balance of the initiative must not only take into account the consumption of the unit or the system improved but even all the changes in power consumption that the initiative has caused in the associated context (cross-media effects). For example, a heat recovery through an exchanger between two streams increases the load losses in the circuits, and therefore the energy consumption of pumps or fans increases. When planning the measure, such components have to be identified and calculated, in order to prevent their occurrence possibly nullifying the operation, causing an increase of the global consumptions: E2 > E1. The absolute ex post consumption may increase rather than decrease when there was a change in the flow of products or services provided after the implementation of the measure; in such a case, a normalization procedure, according to EU Directive 06/32 "whilst ensuring normalization for external conditions that affect energy consumption" [11], is mandatory.

Fourth, the reduction of the energy consumption obtained after the energy-saving operation should remain stable over time. A new and effective measure implemented should become the new benchmark or the new baseline for the same type of energy use and the same final user; since the decision-maker can't accept that, in the midterm, consumptions increase again, nullifying the resources (always scarce and precious) used to implement the measure: a legislative decree, or a company policy, must necessarily induce lasting effects considering the resources invested.

Fifth, the cost-benefit analysis of an energy-saving initiative should provide positive results. The measure must, in fact, be capable to generate, for a number of years established by the decision-maker, a cash flow able to offset the investment necessary for the implementation of the measure itself and to produce an extra advantage, the net present value (NPV). Since the NPV is determined as present money, it can be used to finance further energy-efficient initiatives showing positive NPVs, thus triggering a virtuous spiral. If the energy-saving measure showed a negative NPV instead, the final result would be the loss of money, the value of the total energy saved not being able to pay back the initial investment, making it impossible to trigger any virtuous spiral. This fifth condition moves "energy saving" from the domain of energy to the domain of economy, fixing an inescapable two-way relationship between energy and economy: there is no energy saving if there is no money saving.

In conclusion, "energy saving" can be defined as an operation due to a voluntary and programmable action put in place by the decision-maker, producing a stable, positive, and measurable reduction of energy consumptions between an ex ante and an ex post situation, profitable under an economic point of view.

5. The measures to limit energy consumptions

We have just seen the features for an initiative to be called "energy saving." Let's now see the possible ways to reduce energy consumptions, to discriminate among them only those deserving the "energy-saving" status.

If in a given context it is necessary to submit energy uses to a deep assessment due to consumptions increasing uncontrollably or to associated costs verging on unsustainability[1], possible implementable measures will be the following:

1. Improvement of energy efficiency.
   - In intrinsic way.
     - Via technology.
     - Via management.
   - Joint replacement.
   - Technological standards.
2. Energy waste reduction.
3. Behavioral change.
4. Energy rationing.

We go now to describe respective meanings and operation fields.

5.1 Improvement of energy efficiency

Before getting into the description of this measure, it would be appropriate to clarify the meaning of "energy efficiency," and it would be better to remain in the technical-scientific field, since the concept of "efficiency" comes from the concept of thermodynamic "output." To this purpose it seems consistent what the 2006/32 EC Directive states: energy efficiency is "the ratio of output of (1) performance, (2) service, (3) goods or (4) energy, to the input of energy" (figures added by the author). It is tacit that output and energy input are referred to the same time period, which could be instantaneous (then the efficiency is a ratio between powers) or as long as you like. We can mention the following examples:

1. Performance: for an organization delivering administrative services, energy efficiency could be the ratio between the number of files issued and the energy used as resulting from energy bills.

2. Service: for transport of passengers, it is the ratio between the "number of passengers × kilometers traveled" and the consumption of fuel and/or electricity.

3. Goods: for a paper mill, it is the ratio between the tons of paper produced and the cubic meters of natural gas used in the production process.

4. Energy: for a heat generator, it is the ratio between the thermal energy produced (and fed into the distribution system) and the consumption of primary energy fed to the burner. By definition, the latter type of efficiency coincides

[1] This consideration is not marginal. The entire analysis assumes that an operational context needing energy exists, and it is useful and important. In such a situation, a voluntary interruption of the energy supply is therefore not conceivable, which would seriously prejudice the delivery of the product/service. The article will not therefore consider actions like the interruption of the power supply for a production process during the normal working hours, the turning off of the lights leaving bystanders in the dark, etc.

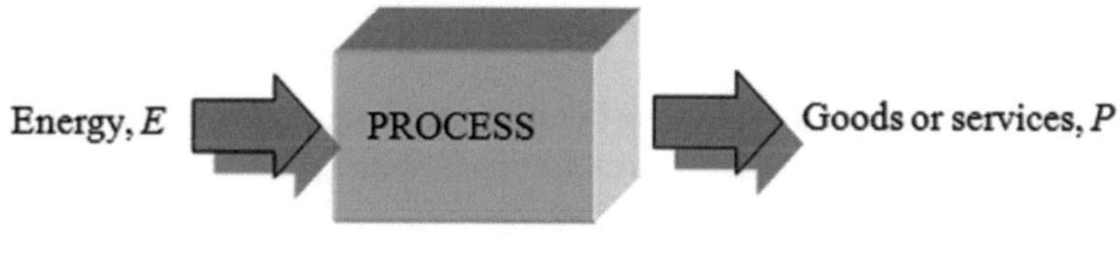

Figure 1.
Energy efficiency as the ratio of goods-services provided and the energy input.

with the first principle of thermodynamic output related to the process taking place in the heat generator.

The measure "improving energy efficiency" is composed of three subsets: the intrinsic improvement, joint replacement, and technological standards.

5.1.1 *Intrinsic improvement of efficiency*

If the efficiency of a process improves thanks to a planned measure, energy saving will be produced, which is therefore the effect of the improvement of efficiency. Let's suppose a given process (see **Figure 1**) which, thanks to the input of energy E, produces the stream of goods or services P. By definition, the energy efficiency of the process is given by.

$$\varepsilon = P/E. \tag{1}$$

If the efficiency of the process improves from ε_1 to ε_2 ($\varepsilon_2 > \varepsilon_1$), the energy savings achievable, R, will be given by

$$R = E_1 - E_2 = P\left(\frac{1}{\varepsilon_1} - \frac{1}{\varepsilon_2}\right) \tag{2}$$

The formula gives essence to the difference between the increase of efficiency and the energy saving: efficiency is a ratio between two quantities that, in the case where the output consists of a supply of energy, becomes a pure number between zero and one. Energy saving is a physical amount of energy instead (measurable in toe, kWh, MJ, etc.), no longer consumed thanks to the increased efficiency. So, increase in efficiency and energy saving are not concepts alternative to one another or possibly overlapping: the first is the cause, the second the effect.

Given the definition of "efficiency" as the ratio $\varepsilon = P/E$, it follows that the inverse of the efficiency is the specific consumption cs = E/P.

5.1.1.1 *Improving efficiency via technology*

The intrinsic increase of the efficiency is obtained via technology when the set of physical equipment driving the process has a better efficiency than the ex ante situation. This occurs, for example, in the presence of IE3 class electric motors instead of IE1-IE2 classes, inverters driving electric motors connected to variable loads, heat exchangers to recover energy from exhaust, steam recompression, turbo expanders in place of lamination valves of gases or steam, etc.

5.1.1.2 *Improving efficiency via management*

Efficiency can be increased by changing not the hardware but (1) the nature of the stream of matter/energy as the process input/output and (2) different management methods. Obviously the current configuration (machineries and the way they are managed) must be an average reference, or baseline, and the possible solution must be a real innovation, not just a realignment with what is already consolidated in the same production sector (we would otherwise fall into the "joint replacement" described in section 5.1.2). Management measures are, for instance:

- The adoption of raw materials with lower energy requirements for a change of state (lower temperatures for melting/boiling) or for pumping (lesser viscosity/density fluids) and use of additives in raw materials conferring the previous properties (e.g., thinners in the production of paper, low-melting additives for the production of glass, etc.)

- The production of lower-energy intensity goods or services (e.g., lighter bricks, lesser-thickness glass containers, avoiding to print documents in favor of dematerialization, etc.)

- The adoption of different management modalities of the process, e.g., "tuning" the different production phases eliminating intermediate stations and queues, using of dedicated software for automation/optimization of process parameters, etc.

5.1.2 Improving efficiency: joint replacement

Joint replacement savings are obtained when a device of a given residual life is replaced by a new one, belonging to the same technological series. The resulting savings are transitional and could even go to zero.

In fact, the efficiency of a new equipment, left to itself, would degrade naturally over time from the rated value ε1 to ε1, f at the end of the use. If at time t1 it was replaced by a new equipment, but belonging to the same technological series of ε1 efficiency, energy savings that could be achieved through the formula (1) would have a purely illusory character (see **Figure 2A**) because, through a normal maintenance, the efficiency would remain at its rated value and consequently the achievable saving would be next to zero (**Figure 2B**).

As we will see in section 5, the energy savings resulting from an intrinsic increase in the energy efficiency (**Figure 2C**) are the only "additional," resulting as marginal quantity of energy corresponding to the value given by (R1), and really saved.

5.1.3 Improving efficiency: measures based on standards

Some decision-making authority may impose minimum performance standards for energy equipment, excluding the circulation in the market of not compliant

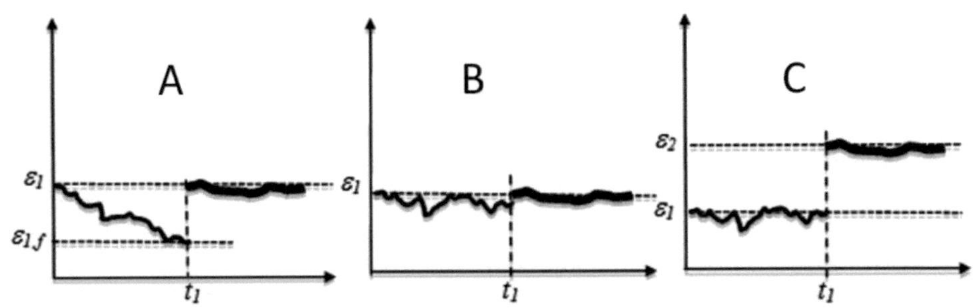

Figure 2.
Type of savings. (A) Illusory savings: when ε1-ε1, f degradation due to lack of maintenance. (B) With proper maintenance saving should be next to zero. (C) The saving is given by an intrinsic improvement of efficiency.

solutions. The obligations imposed by regulations on energy performance in buildings [12, 13] or the minimum standards established by the Ecodesign regulations [14] for different kinds of wide diffusion devices (light bulbs, electric motors, boilers, etc.) belong to this kind of measures.

5.2 Waste energy reduction

The reduction of energy waste comes from the "normal" behavior and it does not need the design of a specific energy project. It simply consists in the realignment to the normal situation at rated consumption, starting from a previous high-energy consumption situation. The energy waste should never be confused with the low efficiency of an appliance, intrinsic or due to wear, which can be increased by substitution with a better device: energy waste is due to negligence, and as such it should not be tolerated and should not be hosted in the country of energetics. If one was currently seeing a "free" degradation of energy, for example, neglected losses of compressed air, steam, and water; lights, printers, and monitors unnecessarily left switched on after office hours; wasted fuel caused by unnecessary travels of the vehicle; etc., the share in case recovered after an initiative of restoration would have the same meaning of the "due" mending of the purse from which a trickle of coins is happening: once the purse is mended, no one should consider as "saved" the money that it is now able to retain.

Metaphors aside, the savings associated with a limitation of wasted energy are illusory: this form of recovery cannot be counted as revenue, and the manager of the local plant is responsible for such a loss.

5.3 Behavioral change

When an energy-saving behavior is adopted, we are dealing with a non-technological measure, related to sociocultural cycles having a complex, long-lasting, unpredictable dynamics difficult to quantify even in ex post conditions. Such an approach exploits the deep motivations of users—citizens—and associates, to a behavior oriented to saving energy, acceptability by the reference community, or the satisfaction of inner instances of public participation, making an active social role evident.

5.4 Energy rationing

Consumptions of energy can be limited by imposing restrictions on energy uses (reducing available streams or the periods of use) or decreasing the quality of performance. This usually happens after serious crises threatening the continuity or security of energy supply or following traumatic increases in energy prices. This kind of tools are typically policies of austerity (i.e., the 1973 Yom Kippur oil crisis) [15, 16] as well as those regulations limiting the temperature in homes.

Given the different measures seen so far to limit consumptions, the following measures belong to the definition of "energy saving" claimed before and substantiate it:

- Intrinsic improvement of efficiency.
- Joint replacement.
- Technological standards.
- Energy rationing whose requirements remain in force in the medium term.

Recovering energy wastes should not strictly be an operation of "energy saving" since it doesn't produce net savings (third condition); on the other hand, the behavioral change and energy rationing measures, the prescriptions of which are in force for a short period of time, do not ensure the stability of savings for years to come (fourth condition).

6. Normalization and additionality

Savings induced by energy rationing policies are not obtained under the same conditions as before, since the end user is encouraged to accept a downgrading of energy performance, for example, using energy only in certain periods of time, having reduced available thermal power, enduring lower temperatures in winter compared to average comfort conditions, higher temperatures—and higher humidity—during summer, etc. Saving energy by increasing efficiency involves, however, the same conditions between ex ante and ex post situations, thus at the same degree days, humidity, services, goods produced, etc. using conventional normalization methods. For example, an operation increasing the efficiency of a process from ε_1 to ε_2, with increased production from P_1 to P_2, would produce, at the same production ex post P_2, an energy saving given by

$$R = E_1 - E_2 = \frac{P_2}{\varepsilon_1} - E_2 = E_2\frac{\varepsilon_2}{\varepsilon_1} - E_2 = E_2\left(\frac{\varepsilon_2}{\varepsilon_1} - 1\right) = E_2\left(\frac{P_2\,E_1}{P_1\,E_2} - 1\right) \quad (3)$$

Since the saving is positive (in respect of the third condition), it follows that $\varepsilon_2 > \varepsilon_1$, which produces a further tautological condition:

$$\frac{P_2\,E_1}{P_1\,E_2} > 1 \Rightarrow \frac{P_2}{P_1} > \frac{E_2}{E_1} \quad (4)$$

stating the following rule: in order to have positive energy savings as a result of an increase in efficiency and production, the relative increase in production must be more than proportional to the relative increase in energy consumption.

The same applies in the event of increase in efficiency with reduction of production.

Of all the ways to save energy, only the intrinsic increase of efficiency has the characteristics of additionality. For instance, as defined in Annex A to document 9/11 of the Italian Authority for Electricity, Gas and Water System, [17] savings are additional when "purified of all not additional energy savings, i.e., those that would have happened anyway as a result of technological, regulatory and market evolution." When, therefore, some savings are achieved because of mandatory legislation, or after installing a "market average" device (even though more efficient than the replaced device), or if providing a service that the market demands with a given minimal performance, all that cannot be considered additional and, in some contexts, may not receive incentives (e.g., in the Italian system of White Certificates).

When a policy boosts improvements in energy efficiency imposing technological standards, the consequent energy savings cannot be considered additional at end-user level, but they can at the national level.

7. The rational use of energy

Let's now jump to the next level and ask ourselves the best definition of "rational use of energy."

For example, the Italian 10/91 law reported the following definition: "set of organic actions intended to promote energy conservation, appropriate use of energy sources (i.e., avoid waste), improvement of technological processes using or transforming energy (i.e., increasing energy efficiency), the development of renewable sources of energy, the replacement of imported energy sources (i.e., the development of indigenous energy sources)." In practice, the rational use of energy is part of the strategy the country adopts to face the energy challenge, whose motivation and urgency comes from the need of security of energy supplies and from the gradual rise of energy prices due to the depletion of fossil fuels. However, the definition of the 10/91 law, in the light of the above, seems inadequate and ambiguous. It considers energy saving and increased efficiency at the same level, officially weaving the two concepts perhaps for the first time. Moreover, the definition does not mention non-technical aspects such as the behavioral change.

We might then reformulate the concept of "rational use of energy" as the "set of organic actions aimed at reducing consumptions through (1) promotion of energy conservation, energy waste reduction, and behavioral change; (2) development and use of renewable energy sources; and (3) development and use of domestic sources of energy," in such a manner of explicitly identifying as many fields of action, conceptually not interfering with one another, each one deserving a specific promotional policy-making use of dedicated tools; in fact it seems logical that a strategy

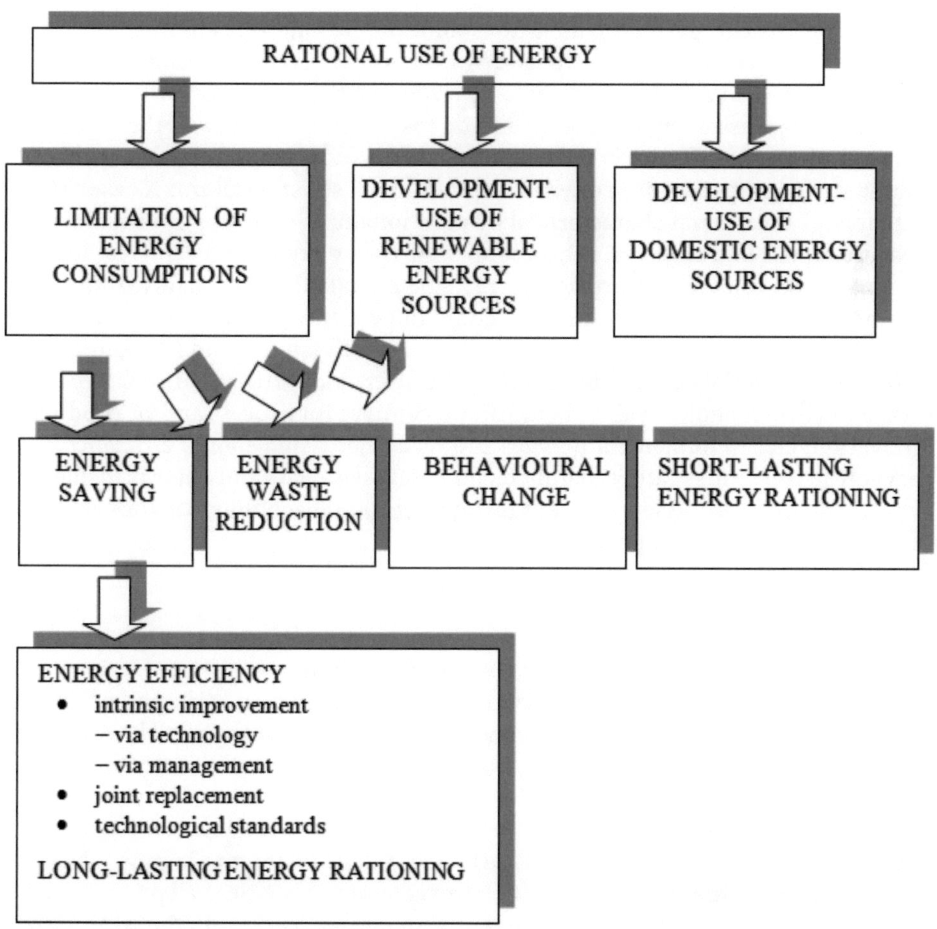

Figure 3.
Rational use of energy: the fields of interest.

for increasing, e.g., intrinsic efficiency, should leverage instruments different than those for behavioral change, joint replacement, etc.

Figure 3 summarizes the proposed definition, depicting the various fields of interest for a policy of rational use of energy.

The diagram shows that energy efficiency is a subset of the family of higher-level "energy saving," an instrument of the class "limitation of consumptions," measure per se of rational use of energy. This should definitely clarify the respective positions and meanings of "efficiency" and "saving."

8. The fields of interest

The proposed subdivision segregates different areas of activity, each one needing dedicated operational tools and skills. Referring to the diagram in **Figure 3**, it is the case at this point to catalog such instruments according to the areas, limiting the analysis to consumption-reducing measures, thus not considering the development and use of renewable and endogenous energy sources.

8.1 "Energy-saving" policies

8.1.1 Intrinsic improvement of energy efficiency

This specific energy-saving measure stands on the planning abilities of the end user—in terms of choice between a number of available technological options or between a range of projects—privileging the solutions with the best efficiency, after assessing them with ad hoc cost-benefit analyses. If the end user has no choice, since the market offers, or the law requires, or customers want only specific equipment or processes, we should not strictly call this measure as "efficiency" for it is not characterized by additionality. Some energy efficiency measures could be as follows: (1) some specific policy granting incentives, for example, based on White Certificates, recognized only if additionality of the energy saved is proven; (2) companies could impose the internal use of efficient equipment and adopt remote and automated control systems; and (3) politics could operate on the demand but even on the offer side, getting industries to produce efficient equipment and granting incentives for process and/or product innovation; etc. To implement policies aiming at increasing energy efficiency, decision-makers must gather mainly technical (researchers and experts in energetics, industrial processes, efficient technologies) and economy-finance skills for the assessment of project profitability.

8.1.2 Joint replacement

This measure should not be subject to specific programs or incentives, since it deals with "average market" practices to be normally adopted at the end of the devices' lifetime and when the efficiency is constantly kept at the rated level thanks to normal maintenance cycles. When, however, the ordinary maintenance is made, but the efficiency of the component degrades necessarily in time, energy saving is obtained by inducing a more frequent renewal of such components (see Appendix). In this regard the Italian Law n. 10/91, Art. One states: "In order to improve the energy transformation processes [...] the provisions of this Title shall promote and encourage [...] a more rapid replacement of systems in particular in the areas with higher energy consumptions" [18]. The implementation of this measure demands technicians and analysts to know the diagrams of decay of the efficiencies in time

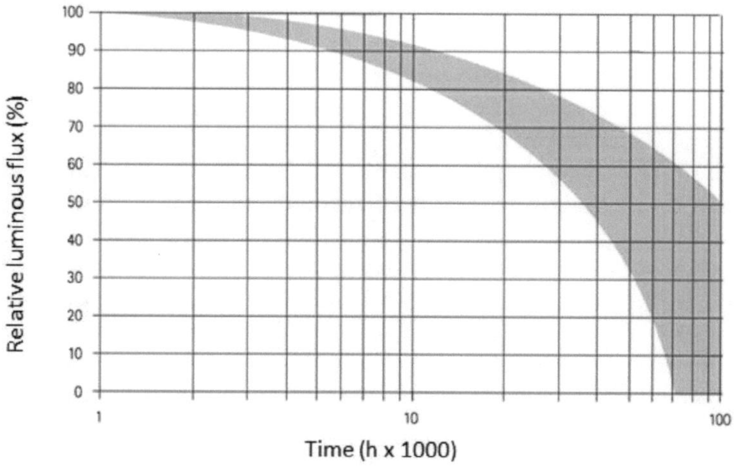

Figure 4.
Decay probability of the luminous flux over time: LED lamps (gray area), source ETAP, LED dossier October 2014.

for the different devices (see, e.g., the decay of the luminous flux for LED lamps [19] in **Figure 4**) and to determine the optimal replacement time through the application of cost-benefit analyses.

8.1.3 Technological standards

Energy saving is achieved by eliminating low-efficiency equipment from the market and allowing only the trade of equipment having an efficiency higher than a given level. European Ecodesign Directives (2005/32/EC, 2009/125/EC, and subsequent regulations) for home and industrial appliances (light sources, digital receivers, electric motors, etc.) leverage this specific item of energy saving. Other measures belong to this category, i.e., imposition of minimum values of transmittance for the building envelope, carrying out of energy audits in large companies, and annual refurbishment of 3% of the buildings of the central public administration (12/27 EED Directive); the requirement for new buildings owned or occupied by public government as of 1 January 2019 has to be "nearly zero energy," while other new buildings will follow the prescription since 1 January 2021 (90/2013 Italian Act).

Such a measure is intended for wide diffusion devices, the number of which constitutes the driver in the country, more than the quantitative increase in efficiency. Efficiency standards are set at EU or Member State level and require the involvement of staff where technical, economic, market knowledge, production processes, protocols for measuring energy consumptions, etc. skills coexist.

8.1.4 Energy rationing

These are measures to be adopted during energy crises involving rapid increases in prices and difficulties in getting supply. When necessary, to achieve drastic reductions of consumptions at whole country, tariffs are increased and energy rationed. For instance, during the 1973–1974 austerity period [20–22], the following measures were adopted in Italy: (1) ban of motorized vehicles (including aircraft and boats) during holidays; (2) end of TV broadcasts at 10:45 pm and evening news moved from 8:30 to 8:00 pm; (3) shops closed at 7:00 pm with the obligation not to hold lightened signs, advertising signs, and shop windows; (4) bars and restaurants closed by h. 00:00 am; (5) cinemas and theaters closed by h. 11:00 pm; (6) immediate increase by

30% in fuel prices; (7) obligation to reduce public lighting by 40%; and (8) reduction of speed on roads to 50 km/h in urban areas, to 100 km/h on country roads, and to 120 km/h on motorways. In other countries the crisis of 1973 was the opportunity to introduce the daylight-savings time [5, 23], while in the USA the bike race Daytona 200 was competed for 180 miles.

In Japan [24, 25], following the tsunami that damaged the Fukushima nuclear power plant (2011), the government implemented a series of measures such as turning off air conditioning and escalators in the subway, turning off the large advertising screens in city centers, decreasing the speed of trains, and reshaping the hours of work, including hours of weekend when the electrical load on the network was lower. Part of these measures were spread by the movement of opinion Setsuden (see **Figure 5**, a poster aimed at private households encouraging energy-saving behavior from unplugging appliances vs. standby mode to turning lights off when not needed and switching to LED lights). Some were removed in late 2011, but part of them was incorporated permanently in the habits of Japanese citizens and companies.

Options foreseen in national laws on energy uses in buildings, for example, limiting the temperature inside homes or in industrial mills during winter and the length of the heating seasons for households, can be considered as belonging to the energy rationing measures.

Planning this instrument primarily requires skills in the field of social psychology, mass psychology, and communication.

8.2 "Energy waste reduction" policies

There are no specific rules in this field because of the difficulty to conduct ex post controls to evaluate the results; however, there are more general measures able to induce at end-user level, among other effects, also an attitude aimed at reducing losses, for instance, a policy accompanying the introduction of energy management systems, possibly compliant with the ISO 50001 Standard [26]: in fact a management system, having among its cornerstones the "continuous improvement," necessarily has to address the problem of energy wastes and the identification of the measures for their reduction. Awareness of the problem of energy waste is also catalyzed by the increase in energy prices. The imposition of limits not to be exceeded for energy-specific consumption at end-user level is aimed at the same

Figure 5.
Banner of the movement of opinion Setsuden (energy saving).

goal. This last measure has never been implemented so far, and, especially in the industry sector, it would be with great difficulty because of the enormous variety and peculiarities of different production cycles and the impossibility to identify and agree with stakeholders the specific energy consumption benchmarks.

Policies to reduce the energy waste, rather than by the central authorities, should be implemented by end users themselves, using local technicians who are familiar with the current consumption profiles, with industry benchmarks to compare energy performances and, of course, with the technical "weaknesses" of their plant. In the end, it is a matter of common sense.

8.3 Behavioral change policies

The behavior of end users regarding the conscious use of energy should be addressed toward proactive attitudes and not toward the trivial (and already due) waste energy reduction. Such a policy should not be therefore aimed at turning the lights off when leaving the living room (or office or department), or shutting the windows when the HVAC plant is switched on, or avoiding the use of compressed air to wipe floors or clothes, or preferring public transport to private cars. Similar fields should already have been fixed by the reasonable user, who knows that unnecessary consumption of electricity or gasoline or methane is a money trickling down; using common sense to prevent similar drippings is enough: there is no need of a law but the law of nature to know that energy—meaning money—shouldn't be wasted. Behavior should change in a proactive way instead, i.e., toward attitudes able to predict the effects of our choices or behaviors about the variable "energy" and then to act accordingly. As a consequence one should consider the energy class label when purchasing a piece of appliance; use elevators and escalators as little as possible and use stairs instead; go walking or cycling for short trips instead of using motorized vehicles; prefer "0 km" products (with same quality and price) and avoid the consumption of fruit and vegetables out of season (greenhouses operating off season must be conditioned, and energy consumptions increase); keep the electrical consumption of the dwelling under control, possibly with automatic meters showing in real time the power requested in that very moment; encourage conference calls or video conferences instead of face-to-face meetings; introduce new summer dress codes in offices, in favor of light and informal clothes (getting rid of jacket-tie suits); learn how to build a solar panel; understand the principles of thermodynamics—stating that (a), once used, the energy can't be created once

	Area of interest	Competences
Energy efficiency improvement	Intrinsic improvement	Energetics, industrial processes, efficient technologies, management systems, economy, finance
	Joint replacement	Technical knowledge of performances of energy components, economy
	Technological standards	Energetics, engineering, economy, knowledge of markets, of productive processes, and of energy consumptions measurement protocols
	Energy rationing	Social psychology, mass psychology, communication
	Energy waste reduction	Common sense
	Behavioral change	Communication, marketing

Table 1.
Behavioral change: skill vs. areas of interest.

again and (b) electricity should never be converted into heat; etc. To achieve such goals, an intense, extensive, and prolonged public information/training plan should be launched, favoring the spread of best practices and involving schools at all levels [9]. Therefore, skills in communication and marketing are mainly needed.

The following table summarizes the most important skills needed in all areas of interest.

9. Conclusions

The definitions and the areas of interest shown so far for the different measures to limit energy consumption have a universal value, not necessarily the one that the different regulations define in terms of "energy saving," "efficiency," "rational use of energy," etc. The distinction proposed between the individual measures by level, areas of interest, and competence requirements can contribute to a better integration of the various regulatory measures and to an optimal identification and customization of planning and implementation tools, while avoiding overlapping and duplications.

The word "energy" has, for the ordinary citizen, a range of meanings: there are vital, moral, mental, psychic, internal, emotional, etc. "energies." These are pure abstract concepts, inhabitants of the world of ideas. The object of policies is, instead, that "energy' is—always abstract but measurable—dealing with the two principles of thermodynamics[2]—and that pragmatically warms us during winter and cools us during summer, making refrigerators and cars run. The policies refer to this energy as a "tangible" energy that burns and gives electric shocks and that can be measured and billed, and that is why the energy efficiency dealt with by the policies should remain restricted to a technical-scientific domain. An invasion of collective areas—behavior modification, increasing awareness, the maturation of a culture or motivational domains, "I feel myself realized" or "If everyone did like me..."—is desirable, but in view of an increased social sensitivity, not of an increase in efficiency, since between the two aspects there is not necessarily a two-way relationship: when sensitivity to energy efficiency is high, the efficiency of the context is normally low (e.g., in energy crisis times), whereas when efficiency is high, sensitivity is low (as in current times). This phase shift may be another form in which we experiment the rebound effect [27–31].

The lack of strict definitions in the field of the rational use of energy has been identified as the main reason for the failure of energy policies aiming at conceptually different goals but with an incoherent bias toward a specific tool to be implemented. When a large-spectrum policy is aimed at different targets, policy-makers should moreover be reminded of the Tinbergen's rule, stating that when trying to achieve multiple economic targets, at least one policy tool for each policy target is needed: the achievement of a target can preclude the achievement of another one.

Another context that the analysis can help to clarify lies in the semantics of the term "energy efficiency." In recent years, the European Union has issued a series of acts in the field of energy conservation: action plans, green papers, directives, framework programs, decisions, etc. These have always invoked the concept of "energy efficiency," and that address was reflected on individual Member States during the adoption of the various acts and directives. This may seem incongruous since, as seen so far, the increase in efficiency is only one of the ways in which energy can be saved.

[2] $E = 0$ and $\Delta S > 0$: in the course of a phenomenon confined in a closed system, the energy E is conserved and entropy S increases.

One answer may lie in semantics.

The main conceptual contender of efficiency is, as seen so far, "energy saving," the instrument that, by virtue of its large domain of intervention, should be invoked in general policies. The concept of "saving" might however not be well received by the end users (nor by the legislature), if recalling—consciously or unconsciously—pauper horizons in which to engineer oneself, having to tighten belts, giving up opportunities, and accepting a lower quality of life. Other terms often used in this context, such as "limitation," "containment," "reduction," and "conservation," remind similar scenarios. Instead the word "efficiency," from the semantic point of view, resonates the positive concept of improvement related to the advent of futuristic technologies: it does not foreshadow some arduous and colorless future overshadowed by attitudes of thrift and saving, but it contains an unlimited, optimistic, and enthusiastic confidence in technology. We do certainly prefer being branded as "efficient" rather than "thrifty." In the collective imagination, efficiency is Thomas Alva Edison, saving is Scrooge. Saving is the bear, efficiency is the bull.

One could say that the choice of the word "efficiency" itself can be seen as the first and most powerful measure of behavioral change that, internalized by European citizens, may have a significant impact in the socioeconomic system.

Appendix: Effects of an increased frequency of replacement of the same efficiency pieces of equipment

Let's naturally decrease the efficiency ε of a given device "1" over time. In **Figure 6** the curve A-B-C represents the evolution in time of the specific power consumption Ps ($= 1/\varepsilon$). During the life (at time H), the device would consume the energy represented by the A-C-H-F area. If device "1" is replaced with one identical "2" at half of its life (time G), the new consumption of the process would be represented by A-B-D-E-H-F area, and the area B-C-E-D would represent the energy saved in this way.

To become an effective energy-saving option, the initiative has to achieve at least the economic parity, so the discounted cash value of the saved energy (black area) in the life of the project must equal the difference between the value of equipment "2" installed at the time G and the residual discounted values of equipment "1" disposed at time G and equipment "2" disposed at time H.

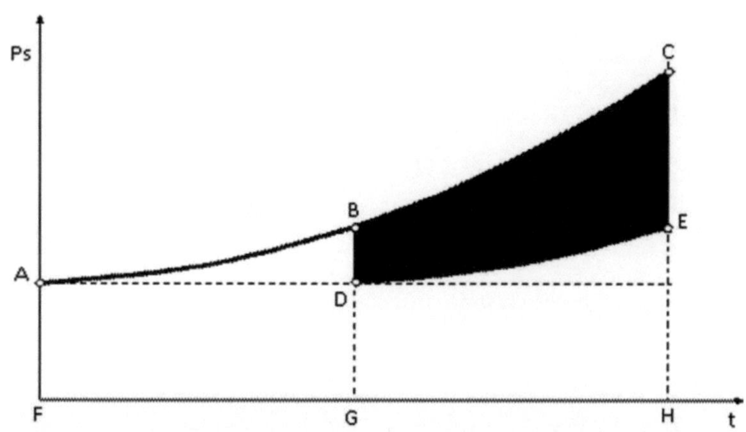

Figure 6.
Power consumption profile of energy-saving solution.

Author details

Nino Di Franco and Mario Jorizzo*
Italian National Agency for New Technologies, Energy and Sustainable Economic Development, Rome, Italy

*Address all correspondence to: mario.jorizzo@enea.it

References

[1] Doms ME, Dunne T. Energy intensity, electricity consumption, and advanced manufacturing-technology usage. Technological Forecasting and Social Change. 1995;**49**:297-310

[2] de Lovinfosse I. How and Why do Policies Change? A Comparison of Renewable Electricity Policies in Belgium, Denmark, Germany, the Netherlands and the UK. Bruxelles: P. I. E. Peter Lang; 2008

[3] European Commission. Energy Efficiency Plan. 2011. Available from: https://eur-lex.europa.eu/legal-content/EN/TXT/?uri=LEGISSUM%3Aen0029

[4] Lovins AB. Energy efficiency, taxonomic overview. In: Encyclopedia of Energy. Vol. 2. San Diego/Oxford, UK: Elsevier; 2004. pp. 383-401. 6 vols

[5] Schipper L. Energy saving in the U.S. and other wealthy countries: Can the momentum be maintained?. In: Tester JW, Wood DO, Ferrari NA, editors. Energy and the Environment in the 21st Century. Cambridge, MA: The MIT Press; 1990:475-484

[6] International Energy Agency. Spreading the Net: The Multiple Benefits of Energy Efficiency Improvements. Insight Series. Paris, France: International Energy Agency; 2012

[7] Niall K, Gouldson A, Barrett J. The rationale for energy efficiency policy: Assessing the recognition of the multiple benefits of energy efficiency retrofit policy. Energy Policy. 2017;**106**:212-221

[8] Kern F, Kivimaa P, Martiskainen M. Policy packaging or policy patching? The development of complex energy efficiency policy mixes. Energy Research and Social Science. 2017;**23**:11-25

[9] Rosenow J, Eyre N. A post mortem of the Green Deal: Austerity, energy efficiency, and failure in British energy policy. Energy Research & Social Science. 2016;**21**:141-144

[10] Decreto-legge 4 giugno 2013, n. 63, recante: Disposizioni urgenti per il recepimento della Direttiva 2010/31/UE. Rome, Italy: G.U. Serie Generale n. 130. del 05-06-2013, (Italian legislation: implementation of EU Directive 2010/31)

[11] Directive 2006/32/EC of the European Parliament and of the Council of 5 April 2006 on energy end-use efficiency and energy services and repealing Council Directive 93/76/EEC. Official Journal of the European Union, L114, 27/4/06

[12] Decreto Legislativo 19 agosto 2005, n. 192, Attuazione della direttiva 2002/91/CE relativa al rendimento energetico nell'edilizia. Rome, Italy: G.U. n. 222. del 23-09-2005, (Italian legislation: implementation of EU Directive 2002/91)

[13] Decreto interministeriale 26 giugno 2015—Adeguamento linee guida nazionali per la certificazione energetica degli edifici. Rome, Italy: G.U. Serie Generale n. 162 del 15-07-2015; Suppl. Ordinario, n. 39 (Italian regulation: national guide lines on building energy efficiency)

[14] Directive 2012/27/EU of the European Parliament and of the Council of 25 October 2012 on energy efficiency, amending Directives 2009/125/EC and 2010/30/EU and repealing Directives 2004/8/EC and 2006/32/EC. Official Journal of the European Union, L 315, 14/11/12

[15] Alpanda S, Peralta-Alva A. Oil crisis, energy-saving technological change and the stock market crash of 1973-1974. Review of Economic Dynamics. 2010;**13**(4)824-842

[16] Chick M. Electricity and Energy Policy in Britain, France and the United States since 1945. Cheltenham: Edward Elgar; 2007

[17] Linee guida per la preparazione, esecuzione e valutazione dei progetti di cui all'articolo 5, comma 1, dei decreti ministeriali 20 luglio 2004 e s.m.i. e per la definizione dei criteri e delle modalità per il rilascio dei titoli di efficienza energetica. Milano, Italy: ARERA, delibera 103/03; (Italian Authority for electricity, gas and water system, Annex A to document 9/11 of Italian Authority for electricity, gas and water system)

[18] Legge 9 gennaio 1991, n. 10 Norme per l'attuazione del Piano energetico nazionale in materia di uso nazionale dell'energia, di risparmio energetico e di sviluppo delle fonti rinnovabili di energia. Rome, Italy: G.U. n. 13. del 16-01-1991. (Italian Law n. 10/91, Art. 1, National energy plan)

[19] ETAP. Lighting with a new light source. In: LED Dossier. 5th ed. Slough, UK: ETAP; 2014

[20] Bernardini C, Salvini G. La Crisi Energetica Nel Mondo e in Italia: da Enrico Fermi e Edoardo Amaldi a Oggi. Bari: Dedalo; 2007. p. 234. (Energy crisis in the world and in Italy)

[21] Caruso E. L'economia italiana negli anni '70-'80, La crisi petrolifera degli anni '70 si abbatte sulle imprese, in «Impresa Oggi», 1999, sezione Economia-Sviluppo economico, Milano, Italy (The Italian economy in the 70s, the oil crisis and the industrial system)

[22] International Energy Agency. Oil Crisis and Climate Challenges, 30 Years of Energy Use in IEA Countries. International Energy Agency: Paris, France; 2004

[23] Tonini A. The EEC commission and european energy policy: A historical appraisal. In: Bardazzi R, Pazienza M, Tonini A, editors. European Energy and Climate Security. Lecture Notes in Energy. Vol. 31. Cham: Springer; 2016

[24] McLellan BC, Zhang QN, Agya U, Hooman F, Ishihara KN. Analysis of Japan's post—Fukushima energy strategy. Energy Strategy Reviews. 2013;**2**(2):190-198

[25] Jupesta J, Suwa A. Sustainable Energy Policy in Japan, Post Fukushima. International Association for Energy Economics, IAEE Forum, Fourth Quarter. Paris, France: IAEE; 2011

[26] ISO 50001—Energy Management. Available from: https://www.iso.org/iso-50001-energy-management.html

[27] European Commission. Action Plan for Energy Efficiency. COM(2006)545 final Brussels, 19/10/06

[28] European Commission. Action Plan to Improve Energy Efficiency in the European Community 2000-2006. Available from: https://ec.europa.eu/energy/en/topics/energy-efficiency [Accessed: April 26, 2000]

[29] European Commission. Green paper on energy efficiency or doing more with less. 2005, COM(2005) 265 final, Brussels, 22.6.2005

[30] European Commission. The Communication from the Commission "Energy efficiency: Achieving the 20% target." 2008, COM(2008) 772 final, Brussels, 13.11.2008

[31] Directive 2012/27/EU of the European Parliament and of the Council of 25 October 2012 on energy efficiency, amending Directives 2009/125/EC and 2010/30/EU and repealing Directives 2004/8/EC and 2006/32/EC90/2013 Italian Act. Final, Brussels: Commission Staff Working Document, Swd; 2016. p. 397. 30.11.2016

Section 2

Microgrids

Chapter 5

Use of Micro-Cogeneration in Microgrids to Support Renewables

Kemal Aygul, Burak Esenboga, Abdurrahman Yavuzdeger, Fırat Ekinci, Tugce Demirdelen and Mehmet Tumay

Abstract

The use of renewable energy sources has experienced great development so as to meet energy demand. With the intention of increasing the utilization of the renewable energy sources near the demand side and compensate the fluctuation of the output power, the use of micro-cogeneration systems with solar (PV) and wind energy overcomes both technical and economic barriers. Micro-cogeneration-based hybrid PV/wind energy system can get stable power output. This new energy model also improves the power quality and significantly reduces the impact of power instability on the power network. In this study, the grid-connected hybrid PV/wind energy-based micro-cogeneration system is modeled and analyzed in detail. In order to test the performance analysis of the system, seven different scenarios are analyzed during the case studies. The analysis results show that the new energy model presents effective solutions to electrical power balance because of its properties such as safety, incombustible structure, and being eco-friendly. It is aimed at providing a broad perspective on the status of optimum design and analysis for the micro-cogeneration-based hybrid PV/wind energy system to the researchers and the application engineers dealing with these issues.

Keywords: eco-friendly energy, micro-cogeneration, wind turbine, photovoltaic, distributed generation, hybrid system

1. Introduction

Due to the increased industrialization, the electricity demand of loads is increasing. As the concerns about environmental pollution increases, the policies for environmental protection have started to become strict. To cope with increasing load demand without violating the environment protection law, the demand for distributed generation (DG) system has increased. Unlike the conventional centralized generation systems, distributed generation does not require long-distance transmission, which emits fewer pollutants. In DG systems the generator, less than 30 MW, is located near to the user side. DG provides the network operator a flexible operation. However, the integration of the DG sources into the grid is a challenging task. It requires domination of the subject of the relationship between distributed sources, feeders, and loads. The examples of the distributed generation sources are as follows: fuel cell, wind turbine, photovoltaic (PV), micro gas turbine, and low-power internal combustion turbine.

Wind turbines and PV panels are DG systems that are the most preferred. But in the recent period, micro gas turbines are often preferred for new applications due to its advantages over other sources such as small size, lightweight, and stable operation capability [1]. Different micro-cogeneration structures were observed when the literature is examined in detail. A novel PI control tuning technique is proposed to refrigerate the hydrogen generation part within the polymer electrolyte membrane fuel cell used in the micro-cogeneration system [2]. A dynamic simulation model of an Ericsson engine is presented for micro-cogeneration systems [3]. Another study in the literature, design, and effectiveness of a highly efficient micro-cogeneration system with a 20 kW prototype fueled by LNG or LPG is presented [4]. A survey and a comparison of basic national testing methods of micro-cogeneration are presented [5]. A sizing optimization procedure is proposed to increase the efficiency of a tubular linear induction generator for free-piston Stirling micro-cogeneration systems [6]. It is presented that the efficiency of micro-cogeneration system including single cylinder diesel engine with an experimental study [7]. Micro-cogeneration in low-energy buildings is proposed by using a load-sharing method. Simulations of two distinct areas in Italy are investigated for thermo-economic efficiency [8]. A micro-cogeneration Stirling unit is investigated for various conditions of the working fluid by experiment and simulation [9]. Efficiency and emission properties of a liquid fuel-fired vugular burner for micro-cogeneration of thermoelectric power are proposed by an experimental study [10]. An extensive literature survey of micro-cogeneration for facilities up to 100 kW consisting of working fluid imaging strategy, elements, expander choice, and detailed properties of industrial and experimental implementation is proposed [11]. Also, a combination of micro-cogeneration and electric vehicle charging systems is analyzed for two distinct areas in Italy with a parametric investigation in simulations [12].

The efficiency of micro-cogeneration systems is investigated with control techniques of Li-ion storage battery by simulations [13]. A high temperature PEM fuel cell based residential micro-cogeneration system is proposed and the detailed mathematical model of whole system is presented [14]. Investigation of the hybrid photovoltaic module-fuel cell combined with microgeneration implementations is presented. Efficiency forecast of the combined system is investigated for discrete climates in Ankara, Turkey [15]. Another fuel cell study, rural micro-cogeneration facility including a high-temperature proton-exchange membrane fuel cell with fuel partialization and power/heat shifting techniques, is presented [16]. A micro-cogeneration system with a solar parabolic collector and direct steam generation is investigated with a prototype [17].

The design model of a new solar micro-cogeneration system with Stirling machine is investigated in terms of efficiency and fuel pass analysis on TRNSYS simulation program for rural areas in Africa [18]. It is proposed that an auto-thermal membrane reformer is combined with a polymer electrolyte membrane fuel cell-based micro-cogeneration system on a prototype [19]. Modeling on TRNSYS simulation and validation by an experimental study of a micro-cogeneration system including an internal combustion engine is presented with over-temperature protection controls [20]. Thermal-economic optimization with generalized pattern search optimization method of a micro-cogeneration system consisting of a parabolic solar collector and Stirling engine is proposed in another micro-cogeneration structure [21].

Systems that use micro gas turbines are also known as micro-cogeneration systems. Micro-cogeneration systems are highly efficient and environment-friendly compared to other conventional energy sources shown in **Figure 1**, because they produce second energy by using exhaust gases. Exhaust gases from the gas turbine are used in a waste heat boiler to obtain high-efficiency heat energy. Therefore, the waste heat is converted to usable energy in the micro-cogeneration system. Micro-cogeneration systems with gas engine also supply economical and eco-friendly concentrated heat and power. Eco-friendly micro-cogeneration power plants with combined heat and power enable economical and energy-efficient power production.

Cogeneration systems are 33% more efficient than coal that generates the same amount of heat and energy [22]. This paper analyzes the electrical modeling and performance investigation of a micro-cogeneration system in a microgrid to support renewables. This system comprises 30 kVA micro-cogeneration system, 10 kVA wind power station, 10 kVA photovoltaic power station, and local electrical loads. The system is designed by using real-time data. Firstly, PV and wind power station are modeled and simulated with the help of the system parameters. Then, a micro-cogeneration system suitable to the hybrid system has been designed by calculating the optimum efficiency.

Due to the limitations of the present studies in literature, the aim of this paper is:

- To demonstrate the mathematical model of the micro-cogeneration-based hybrid PV/wind energy system in detail.

- To test the performance analysis of the system, seven different scenarios are analyzed during the case studies firstly.

- To compare the case studies and investigate the performance.

- To get optimum performance for the implementation of the micro-cogeneration-based hybrid PV/wind energy system.

This paper primarily focuses on the aforesaid four aspects of the proposed system.

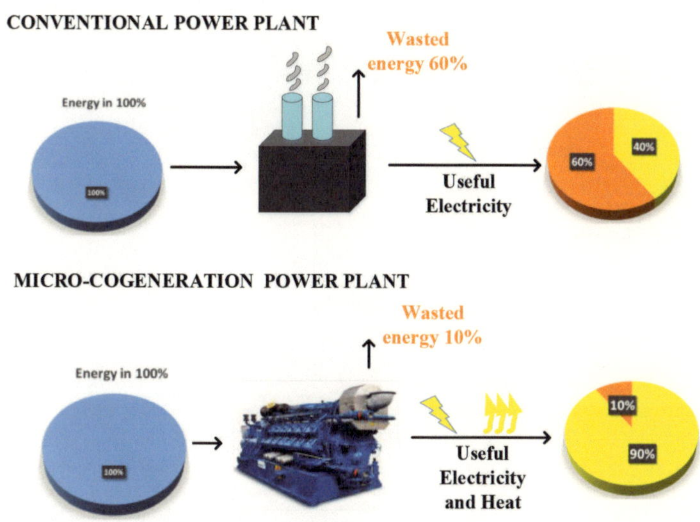

Figure 1.
Energy saving in micro-cogeneration power plant.

2. Mathematical modeling and control of the proposed system

The proposed system consists of two parts: PV and wind system. These systems are examined in detail. The block diagram of the proposed system is shown in **Figure 2**. The control mechanism of **Figure 2** is explained in detail in the following section. The system consists of PV and wind power station, micro-cogeneration system, and the electrical grid. This integrated system feeds the dynamic electrical loads.

2.1 Mathematical modeling and control of micro-cogeneration system

Temperature control, speed control, fuel control, turbine dynamics, and acceleration control block are included in the micro-cogeneration system. The speed control provides to correct the speed error between the reference speed and the rotor speed of the permanent magnet generator system. It is the main control tool for microturbine under partial load conditions. Speed control modeling is done by using a lead-lag transfer function or by a PID controller [23]. Speed control for the micro-cogeneration system is shown in **Figure 3**.

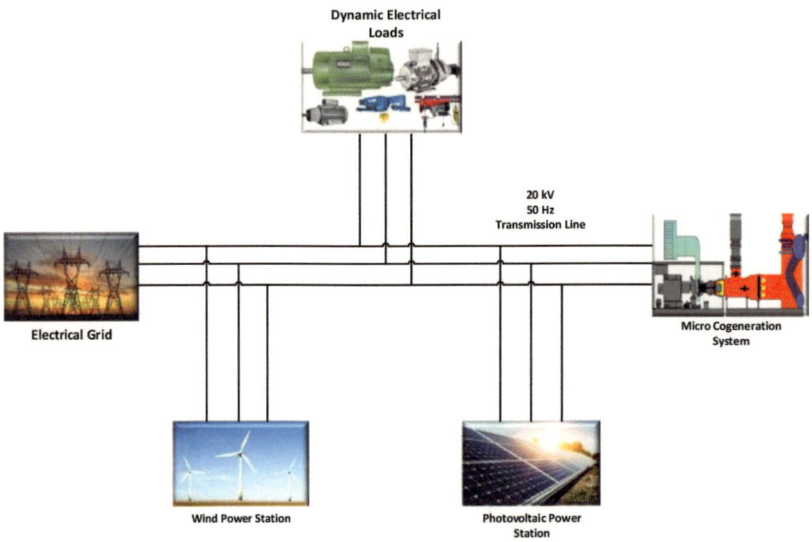

Figure 2.
The block diagram of the proposed system.

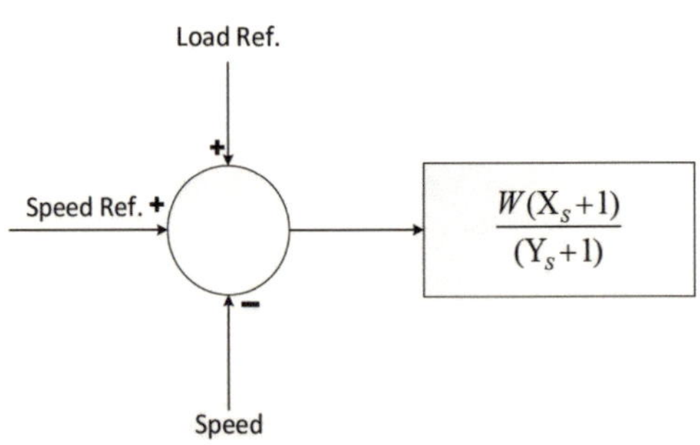

Figure 3.
Speed control for the micro-cogeneration system.

W is the controller gain, X and Y are the governor lead and lag time constant, and Z is a constant representing the governor mode (droop or isochronous). X, Y, and Z can be adjusted so that the governor can act with droop or as an isochronous governor. Acceleration control allows limiting the rate of the rotor acceleration prior to reaching operating speed during turbine start-up. The fuel system consists of the fuel valve and actuator. The fuel system control is provided by the actuator and of the valve positioner shown in **Figure 4**.

V_{ce} is the fuel flow control and the valve positioner transfer function is

$$E_1 = \frac{a}{(\mathrm{b}_s + \mathrm{c})} \tag{1}$$

and the fuel system actuator transfer function is

$$E_2 = \frac{1}{(\mathrm{TF}_s + 1)} \tag{2}$$

In Eq. (1) and (2), a is the valve positioner (fuel system actuator) gain, b and TF are the valve positioner and fuel system actuator time constants, c is a constant, E1 is the input and output of the valve positioner, and E2 is the fuel demand signal in pu.

Temperature control allows limiting the gas turbine output power at a predetermined firing temperature, independent of variation in ambient temperature or fuel characteristics. The fuel burned in the burner causes the movement of the turbine (torque) and the exhaust gas temperature. The exhaust temperature is measured using a series of thermocouples incorporating radiation shields as shown in **Figure 5** [24].

0.8 and 0.2 values are constants associated with the radiation shield, and 3.3 value is the time constant associated with a temperature controller. TT is the temperature controller integration rate, and 15 and 2.5 values are time constants associated with the radiation shield and thermocouple, respectively.

Permanent magnet generators are superior alternatives to conventional induction motors that can be combined with turbines. The main advantages of PMSG are considerably significant: higher operational reliability, higher

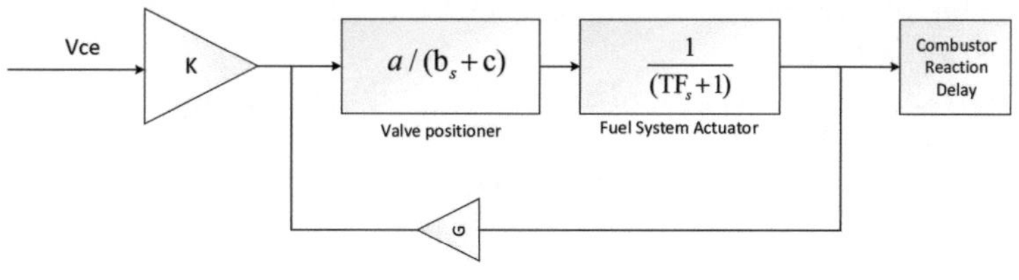

Figure 4.
Fuel system control for the micro-cogeneration system.

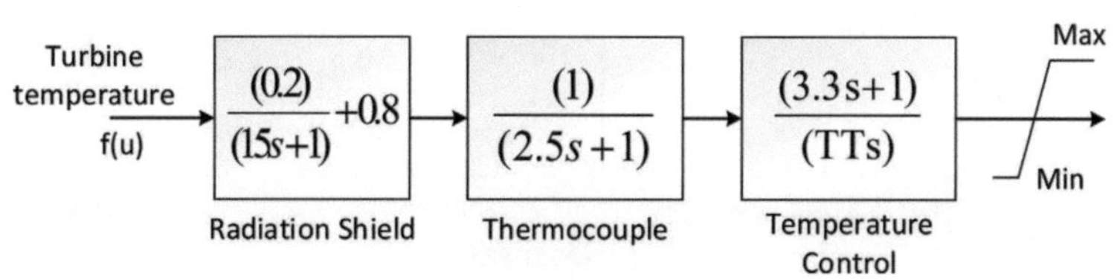

Figure 5.
Temperature controller.

efficiency, and very small energy loss. The microturbine generates electrical energy through a high-speed permanent magnet generator driven directly by the turbo compressor shaft. Totally, 30 kVA electricity is produced by the proposed micro-cogeneration system. The model adopted for the generator is a 2-pole permanent magnet machine generator with a non-salient rotor. At 50 hertz (3000 rpm), the machine output power is 30 kW, and its terminal line-to-line voltage is 380 V.

dw (in pu) is angular velocity, m is speed signal, and Pm is specified as shaft mechanical torque value shown in **Figure 6**.

The electrical and mechanical equations of permanent magnet machine expressed in rotor reference frame dq are as follows [25]:

Electrical equations:

$$\frac{d}{dt}i_d = \frac{1}{L_d}v_d - \frac{R}{L_d}i_d + \frac{L_q}{L_d}pw_r i_q \tag{3}$$

$$\frac{d}{dt}i_q = \frac{1}{L_q}v_q - \frac{R}{L_q}i_q - \frac{L_d}{L_q}pw_r i_d - \frac{\lambda p w_r}{L_q} \tag{4}$$

$$T_e = 1.5p\left(\lambda i_q + \left(L_d - L_q\right)i_d i_q\right) \tag{5}$$

Mechanical equations:

$$\frac{d}{dt}w_r = \frac{1}{J}(T_e - Fw_r - T_M) \tag{6}$$

$$\frac{d\theta}{dt} = w_r \tag{7}$$

2.2 Mathematical modeling and control of PV system

PV power plant is modeled by using five parallel strings and six series-connected modules per string in order to obtain 10 kWP solar PV power. Solar PV module data is indicated in **Table 1**.

One basic solar cell equivalent circuit model in common use is the single-diode model, which is derived from physical principles. The equivalent circuit of a solar PV cell can be represented with a current source which is connected parallel with a diode as shown in **Figure** 7 [26].

This equivalent circuit is formulated using Kirchhoff's current law for current:

$$I = I_g - I_d - I_{sh} \tag{8}$$

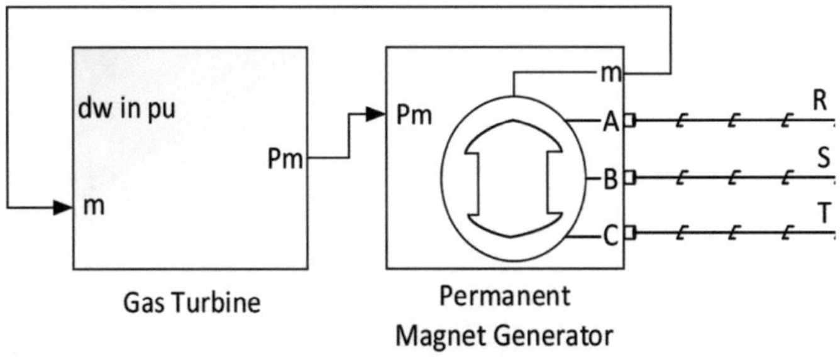

Figure 6.
Permanent magnet generator model implemented in SimPowerSystems.

Maximum power (W)	414.801	Cells per module (Ncell)	128
Open circuit voltage Voc (V)	85.3	Short-circuit current Isc (A)	6.09
The voltage at maximum power point Vmp (V)	72.9	Current at maximum power point Imp (A)	5.69

Table 1.
PV module parameters.

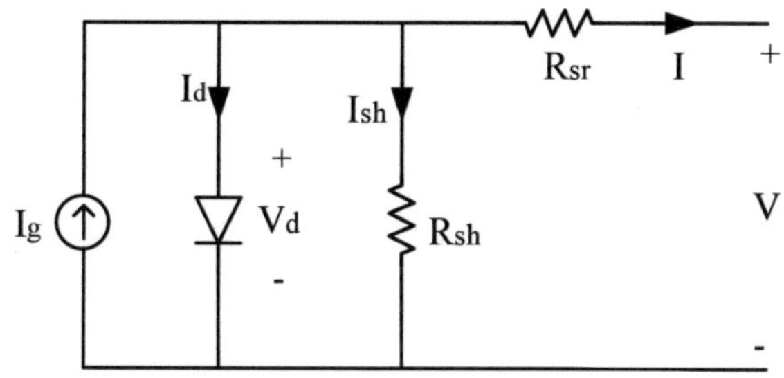

Figure 7.
Equivalent circuit of solar PV cell.

where I_g represents the light-generated current in the cell, I_d represents the voltage-dependent current lost to recombination, and I_{sh} represents the current loss due to shunt resistances. The substitution of the related expressions for the diode current I_d and the shunt branch current I_{sh} is given in Eq. (4):

$$I = I_g - I_0\left[\exp\left(\frac{V + IR_{sr}}{nkT_c/q}\right) - 1\right] - \left(\frac{V + IR_{sr}}{R_{sh}}\right) \tag{9}$$

where n is the diode ideality, I_0 is the saturation current, k is Boltzmann's constant (1.381×10^{-23} J/K), and q is the elementary charge (1.602×10^{-19}C). A group of single cells can be connected either series or parallel combination. By connecting solar cells, PV module can be created. Similarly, by connecting PV modules, a PV array can be created. Power calculation of a solar PV cell is obtained by the equation:

$$P_{pv(t)} = VxI \tag{10}$$

where $P_{pv(t)}$ is solar PV cell DC power, V is solar cell or PV array voltage, and I is current flowing from solar PV cell.

If the generated power on the PV module does not match the load power, there will be an efficiency loss. To overcome this problem, a PV panel must operate at its maximum power point. The variations in solar radiation and temperature affect the maximum power point. In order to prevent losses due to operating point, a maximum power point tracker should be used in PV power systems. For the proposed system Perturb and Observe (P&O) algorithm is used to track the maximum power point. The purpose of the algorithm is to find VMPP and IMPP points that PV system delivers the maximum point. The algorithm is based on a periodic increase and decrease in PV voltage. After an increase/decrease, the system checks the output power. According to the tendency of the output power, the tracker decides if the next voltage perturbation will be in the same way or the opposite way [27]. The flowchart of the P&O algorithm is shown in **Figure 8**.

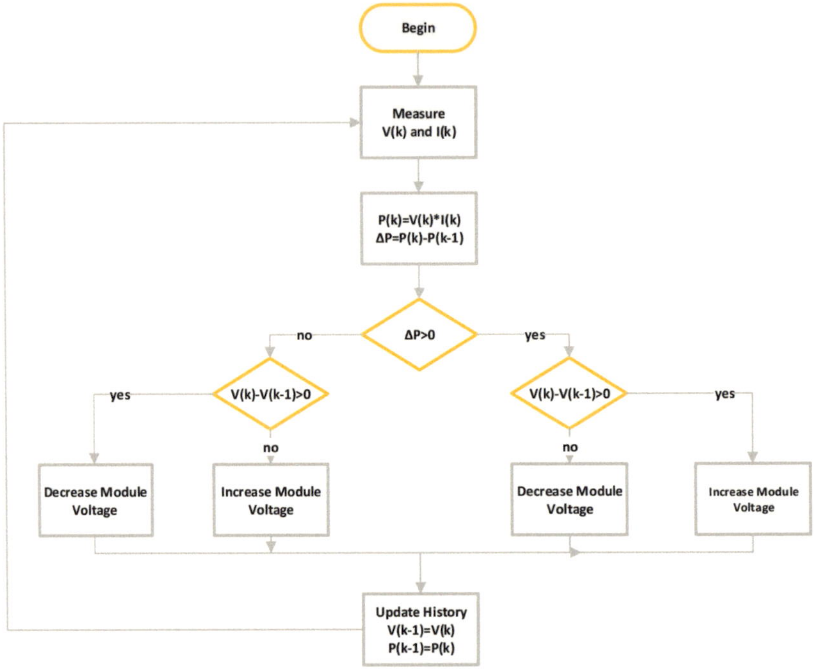

Figure 8.
The flowchart of the P&O algorithm [28].

2.3 Mathematical modeling and control of wind energy conversion system

The proposed wind turbine consists of a rotor mounted to a nacelle and a tower with two or more blades mechanically connected to a wind turbine electric generator. Wind passes over the blades of the wind turbine. Wind power extracted from wind is expressed in Eq. (6) [29]:

$$P_m = \frac{1}{2}.C_p.(\lambda.\beta).\rho.\mathrm{A.v}^3 \tag{11}$$

where $C_p(\lambda\beta)$ is the power coefficient, λ is named as tip speed, β (degree) is the pitch angle of the rotor blades, A is swept area (πr^2), ρ is air density (1.25 kgm^2), and v is named wind speed (m/sn).

The rotating blades turn a shaft that goes into a gearbox in nacelle. The blades of the wind turbine create kinetic energy given Eq. (7):

$$E = 0.5 \ \mathrm{mv}^2 \tag{12}$$

where m is the air mass and v is the wind speed. Air power is obtained by the time derivative of kinetic energy. Air power is given in the following equation:

$$P_w = \frac{d(0,5.\mathrm{m.v}^2)}{dt} \tag{13}$$

where m is the mass flow rate per second and air power is obtained from Eq. (9):

$$P_w = 0,5.m.\rho.\mathrm{A.v}^2 \tag{14}$$

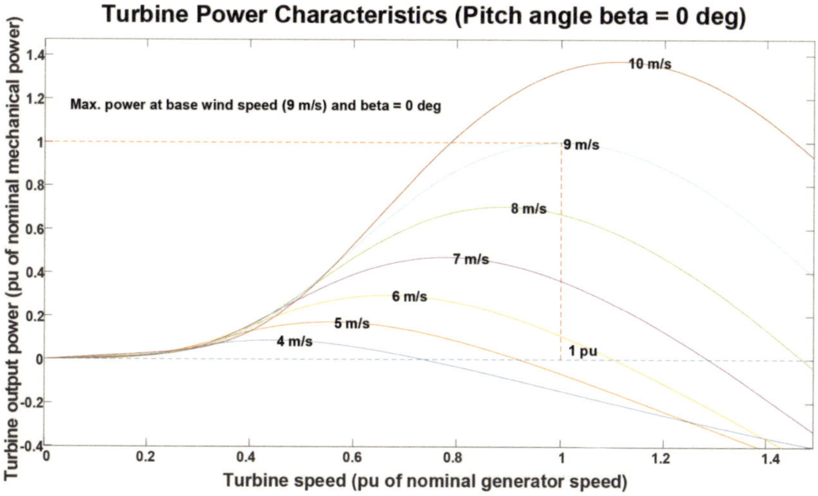

Figure 9.
Power characteristics of the wind turbine.

Blade power resulted by wind effect is given by Eq. (10):

$$P_{blade} = C_p(\lambda.\beta).0,5.m.\rho.A.v^3 \tag{15}$$

Tip speed and rotor torque theoretical parameters are given in Eqs. (11) and (12):

$$\lambda = \frac{w_m.R}{v} \tag{16}$$

$$T_{wr} = \frac{P_{blade}}{w_m} = \frac{C_p.(\lambda.\beta).\rho.A.v^3}{2.w_m} \tag{17}$$

where T_{wr} is the rotor torque and w_m is the angular velocity of the rotor.

The gearbox in the mechanical assembly named as drive train mechanism provides to transform slower rotational speeds of the wind turbine to higher rotational speeds on the wind turbine electric generator. The rotation of the electric generator's shaft generates wind turbine power.

The power coefficient of the wind turbine, therefore, the generated power can be controlled by changing the pitch angle (β). **Figure 9** shows that as the wind

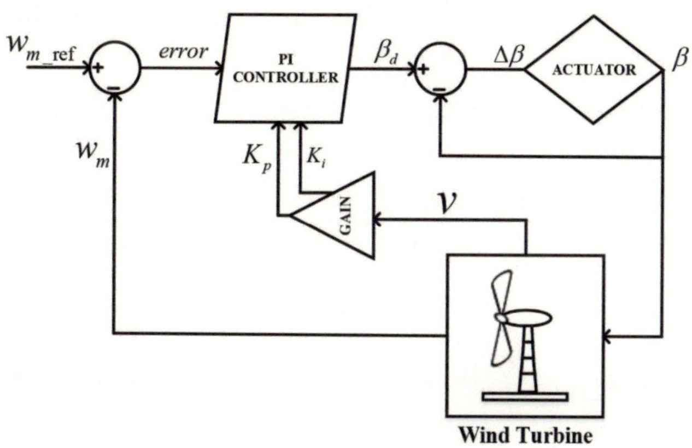

Figure 10.
Block diagram of pitch angle PI controller [31].

speed increases, the amount of power generated by the turbine also increases. According to Eq. (11), further increase in wind speed will result in a further increase in turbine power. But this will also bring a random surge of power. In order to prevent this condition, a control system must keep the power at a constant level. PI controller is used for this purpose in the proposed system [30].

Figure 10 shows the block diagram of the pitch angle PI controller. The pitch angle controller checks the speed and compares it with a reference speed. If the speed is greater than the reference speed, the controller changes the pitch angle in order to approach the rated speed [31].

3. Performance analysis of proposed system and controllers

In this section, the performance evaluation of the proposed system and controllers will be evaluated. The proposed system is modeled using MATLAB/SIMULINK. To test the performance of the system, seven different simulation scenarios are created. In the simulation study, parameters such as wind speed, solar radiation, temperature, and electrical load demand are changed and the resulting graphs of active power, voltage, and currents for each element were taken. The summary of the scenarios can be seen in **Table 2**.

3.1 Case 1

In this case, the parameters such as wind speed, solar radiation, temperature, electrical load demand and the fuel flow of the micro gas turbine are assumed steady during simulation time. The active power flow in the proposed system can be seen in **Figure 11**.

3.2 Case 2

In this case, only solar radiation and temperature on the PV module are variable. This variation affects the output power of the PV station, hence, the grid power. The resulting active power can be seen in **Figure 12**. RMS voltages and currents of the grid and the PV station can be seen from **Figures 13** and **14**, respectively; **Figure 15** shows solar radiation and temperature during simulation of this case.

In this scenario, it is observed that the variations in solar radiation and temperature make the output power of the PV plant variable. Hence, the balance in case 1 is no longer observable. The missing power is supplied by the electrical grid.

Cases	PV system	WECS	Load	Micro-cogeneration system
Case 1	Steady	Steady	Steady	Steady
Case 2	Variable	Steady	Steady	Steady
Case 3	Variable	Variable	Steady	Steady
Case 4	Variable	Steady	Variable	Steady
Case 5	Steady	Variable	Variable	Steady
Case 6	Steady	Steady	Variable	Steady
Case 7	Variable	Variable	Variable	Steady

Table 2.
The summary of the scenarios.

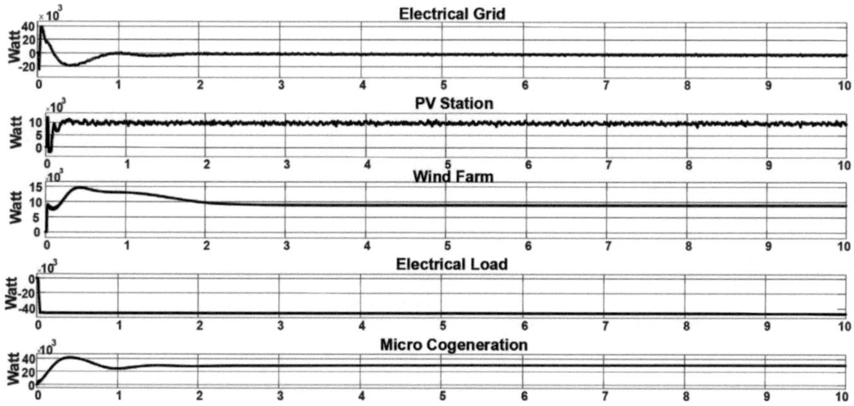

Figure 11.
The graph of active power of the elements in proposed system for case 1 (wind speed 10 m/s, solar irradiance 1000 W/m², temperature 25°C).

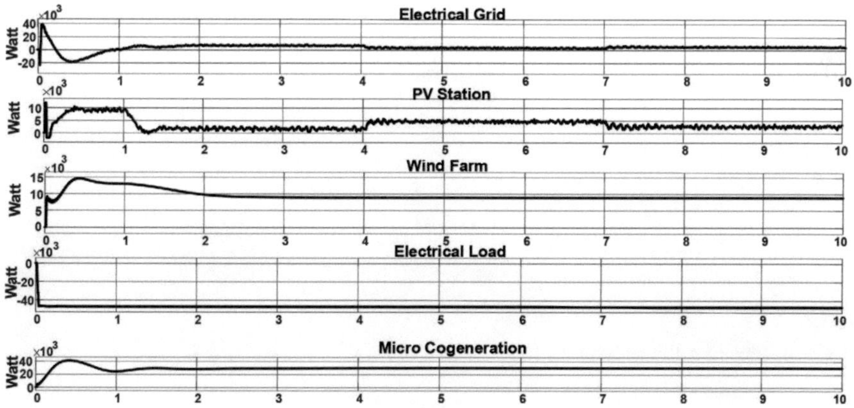

Figure 12.
The graph of active power of the elements in proposed system for case 2 (wind speed is 10 m/s).

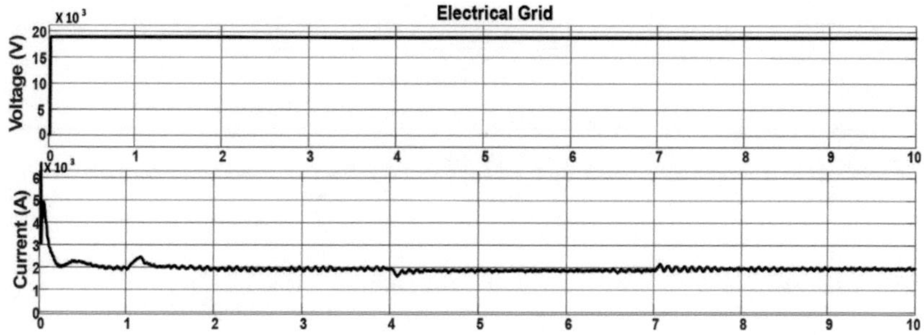

Figure 13.
RMS voltage and RMS current of electrical grid for case 2.

3.3 Case 3

In this case wind speed, solar radiation, and temperature are variable. This variation affects the output power of the wind farm and PV station and the power fed to the electrical grid. The resulting active power graph can be seen in **Figure 16**. RMS voltages and currents of the electrical grid, PV station, and wind farm can be

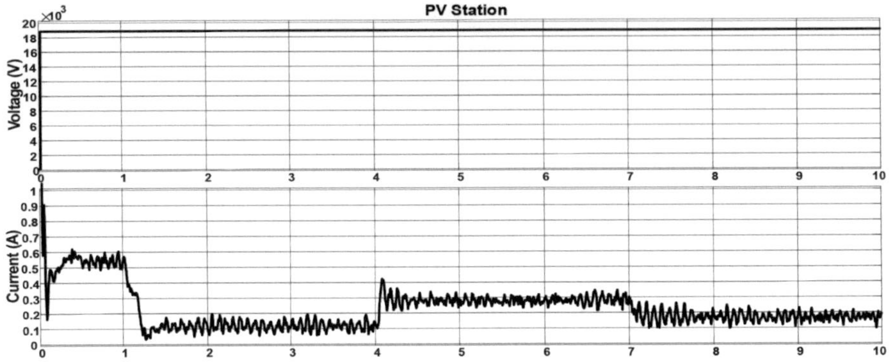

Figure 14.
RMS voltage and RMS current of PV station for case 2.

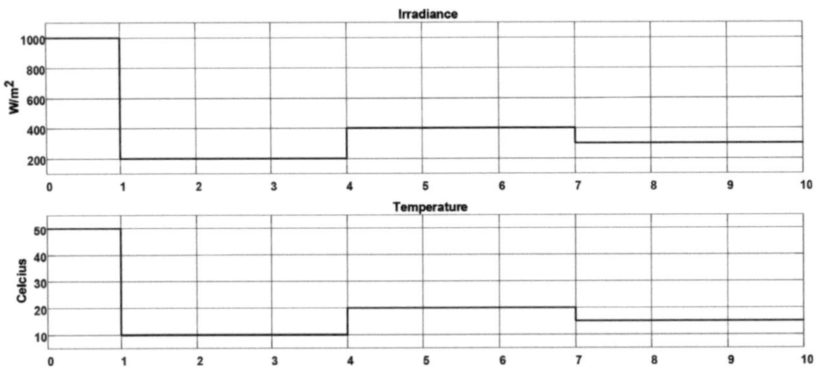

Figure 15.
Graph of solar radiation and temperature for case 2.

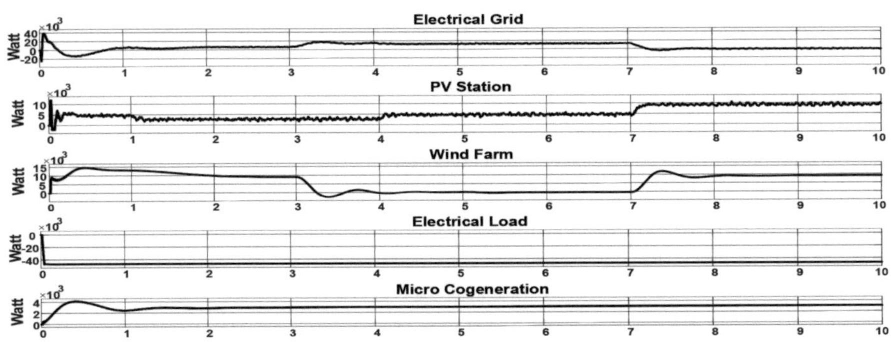

Figure 16.
The graph of active power of the elements in proposed system for case 3 (wind speed is 10 m/s).

seen from **Figures 17–19**. The graph of wind speed, solar radiation, and temperature can be seen from **Figures 20** and **21**.

In this scenario in addition to case 2, the wind speed was also variable. The missing power is supplied by the electrical grid. Variations in wind speed caused better observability of variations in grid power.

3.4 Case 4

In this case, electrical load demand, solar radiation, and temperature are variable. This variation affects the output power PV station and the power fed to/from

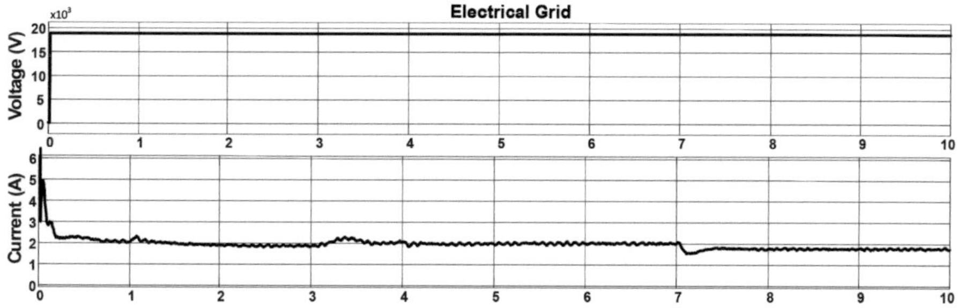

Figure 17.
RMS voltage and RMS current of electrical grid for case 3.

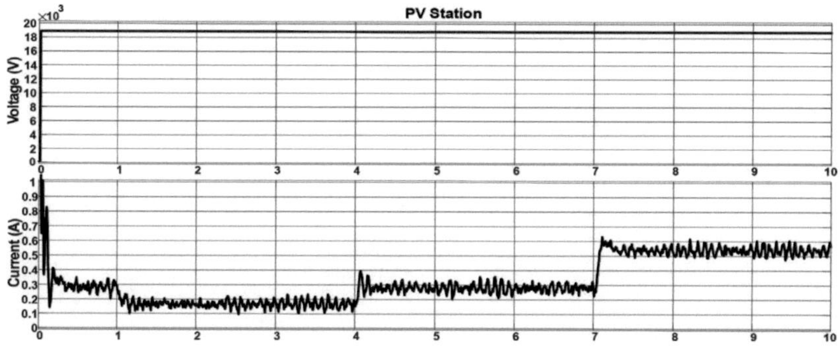

Figure 18.
RMS voltage and RMS current of PV station for case 3.

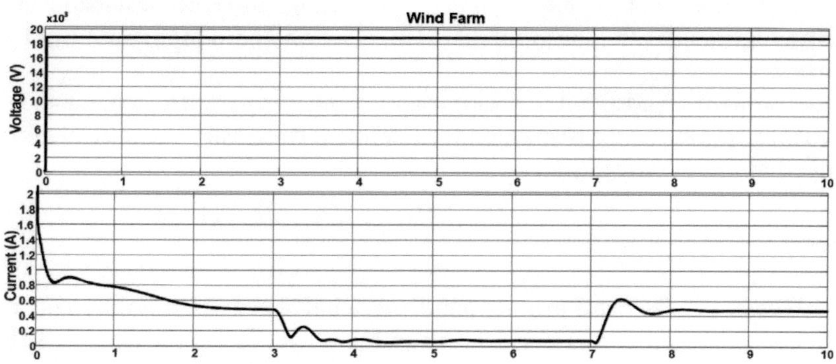

Figure 19.
RMS voltage and RMS current of wind farm fort case 3.

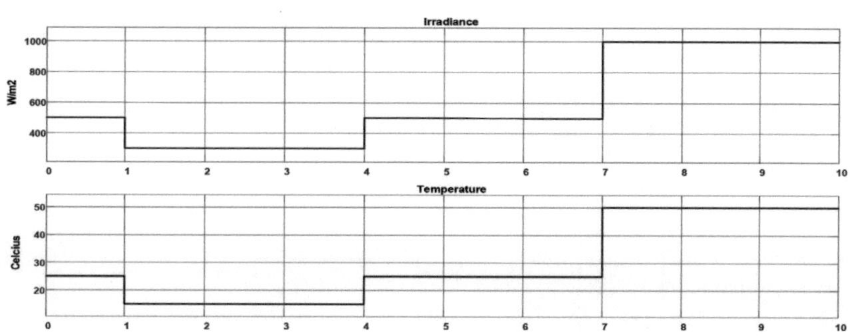

Figure 20.
Graph of solar radiation and temperature for case 3.

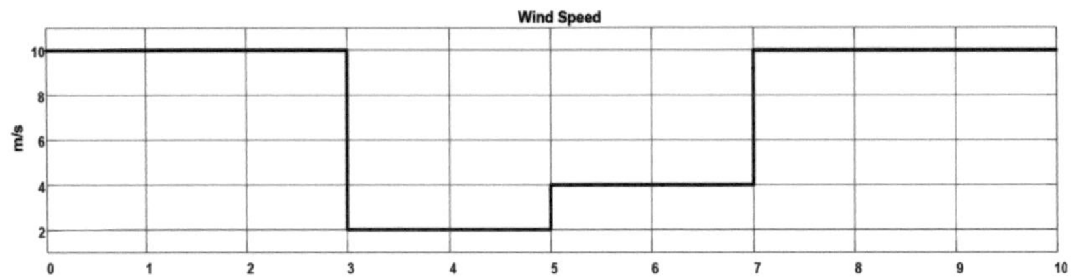

Figure 21.
Graph of wind speed for case 3.

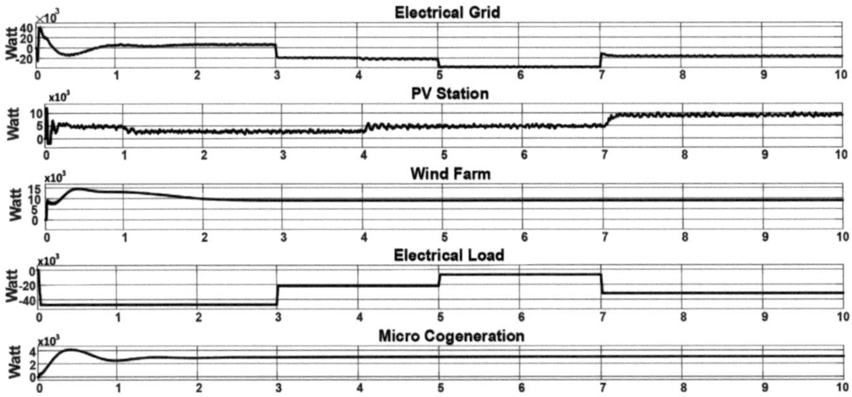

Figure 22.
The graph of active power of the elements in proposed system for case 4.

the electrical grid. The resulting active power graph can be seen in **Figure 22**. RMS voltages and currents of the electrical grid, PV station, and electrical load can be seen from **Figures 23–25**. The graph of solar radiation and temperature can be seen in **Figure 26**.

In this scenario in addition to solar radiation and temperature, load demand was also variable. It is observed that when load demand decreases, the excessive power is supplied to the electrical grid.

3.5 Case 5

In this case, the electrical load demand and wind speed are variable. This variation affects the output power wind farm and the power fed to the electrical grid. The resulting active power graph can be seen in **Figure 27**. RMS voltages and

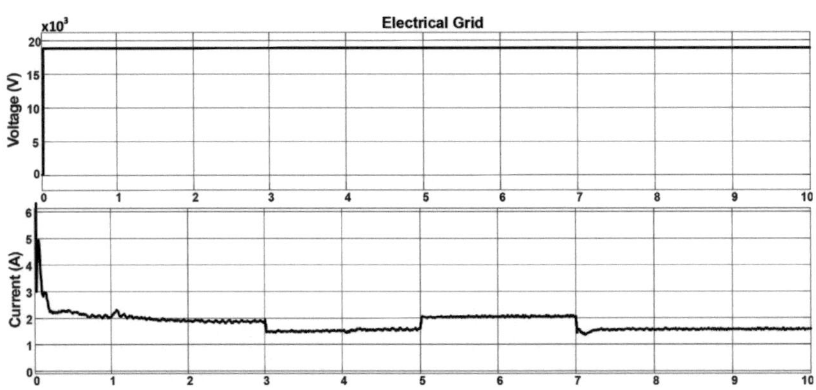

Figure 23.
RMS voltage and RMS current of electrical grid for case 4.

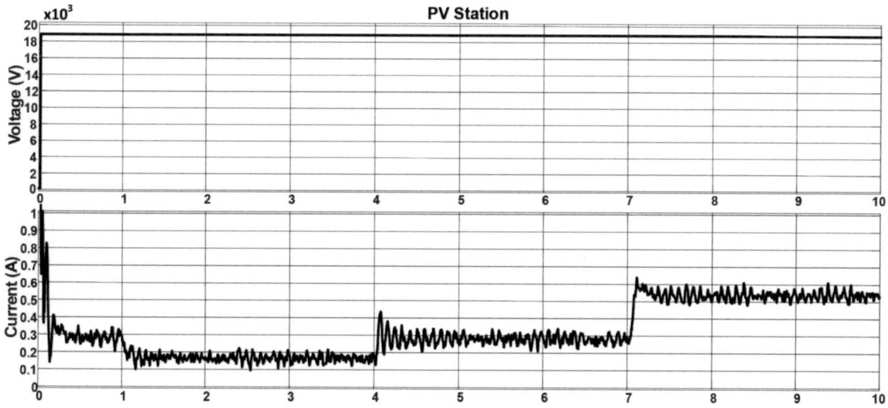

Figure 24.
RMS voltage and RMS current of PV station for case 4.

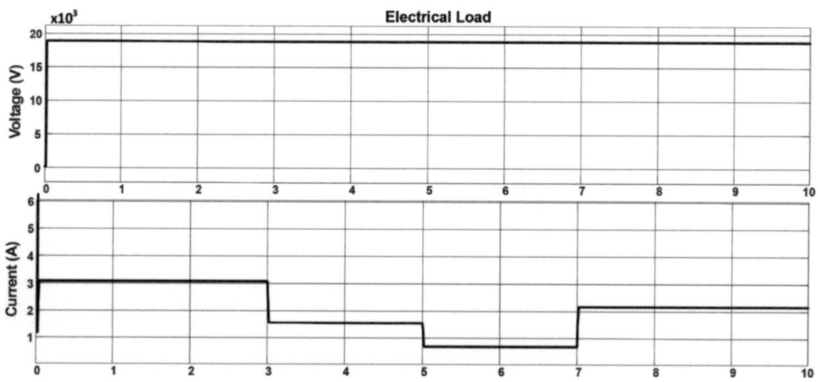

Figure 25.
RMS voltage and RMS current of electrical load for case 4.

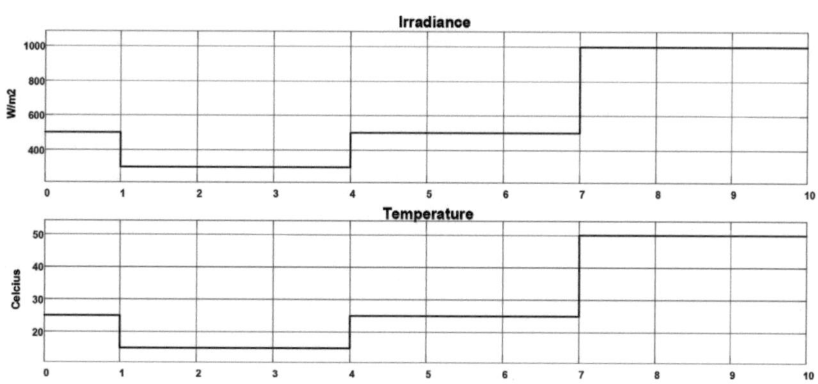

Figure 26.
Graph of solar radiation and temperature for case 4 (wind speed 10 m/s).

currents of the electrical grid, wind farm, and electrical load can be seen from **Figures 28–30**. The graph of wind speed can be seen in **Figure 31**.

In this scenario, the wind power generation and the load demand were variable. It is observed that when the wind power generation is decreased and the load demand is increased, the excessive load demand is supplied from the electrical grid.

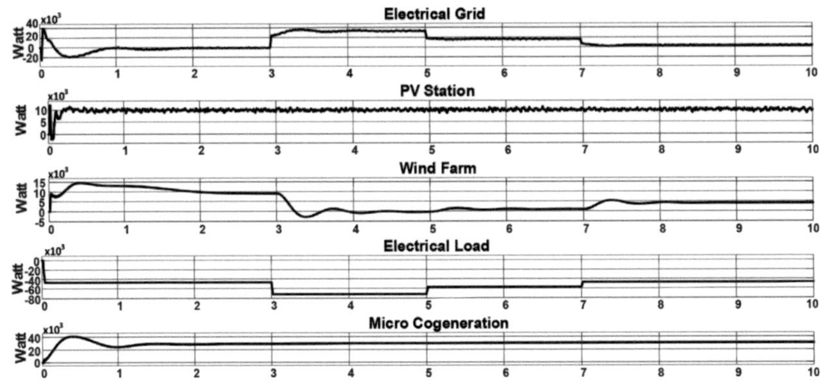

Figure 27.
The graph of active power of the elements in proposed system for case 5.

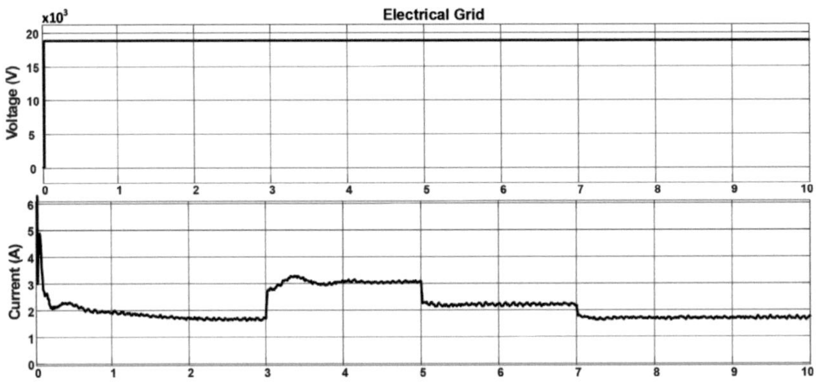

Figure 28.
RMS voltage and RMS current of electrical grid for case 5.

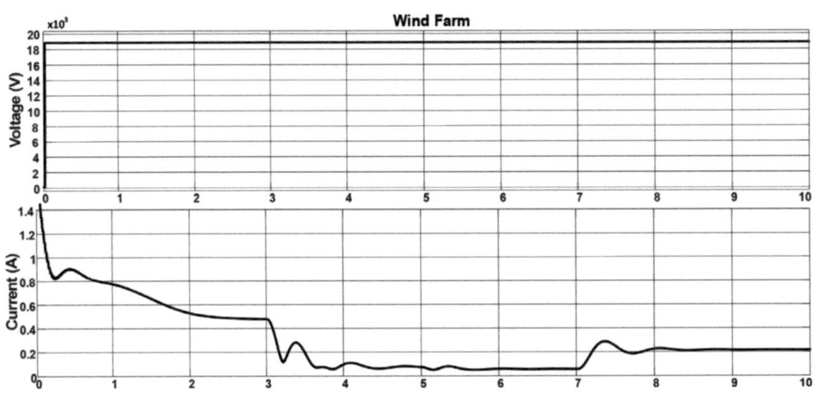

Figure 29.
RMS voltage and RMS current of wind farm for case 5.

3.6 Case 6

In this case, electrical load demand is variable. This variation affects the power fed to the electrical grid. The resulting active power graph can be seen in **Figure 32**. RMS voltages and currents of the electrical grid and electrical load can be seen from **Figures 33** and **34**.

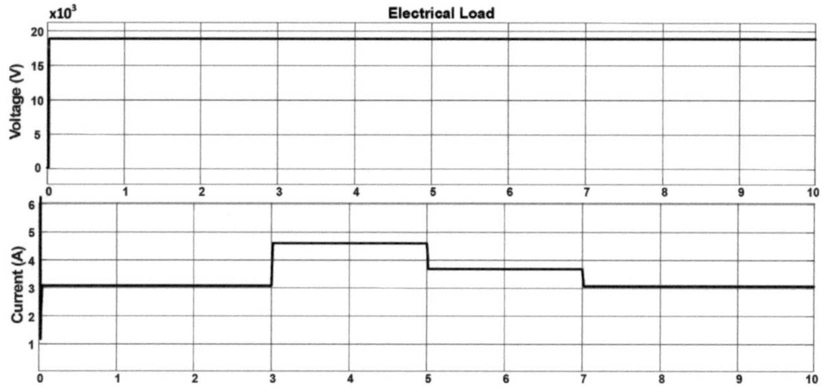

Figure 30.
RMS voltage and RMS current of electrical load for case 5.

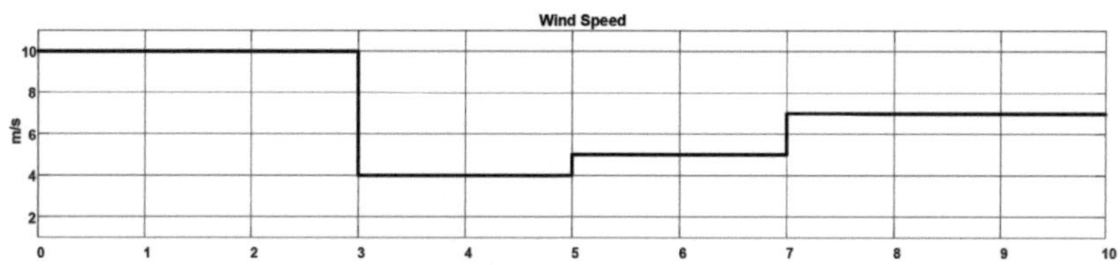

Figure 31.
Graph of wind speed for case 5.

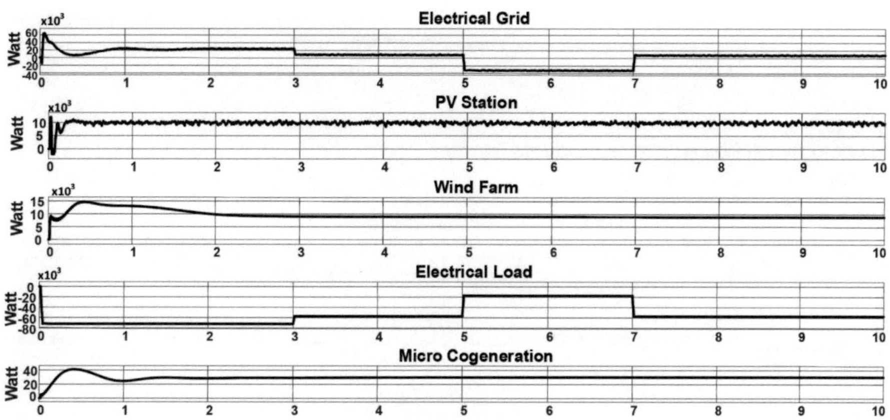

Figure 32.
The graph of active power of the elements in proposed system for case 6.

In this scenario, the only variable parameter was load demand, and both the supply of loads from the electrical grid and the transfer of excess power to the electrical grid have been observed. It is seen that the variations in load demand caused the same amount but reversed variation in the grid power.

3.7 Case 7

In this case, the electrical load demand, wind speed, solar radiation, and temperature are variable. This variation affects the output power of a wind farm, PV station, and the power fed to the electrical grid. The resulting active power graphs can be seen in **Figure 35**. RMS voltages and currents of all the elements can be seen

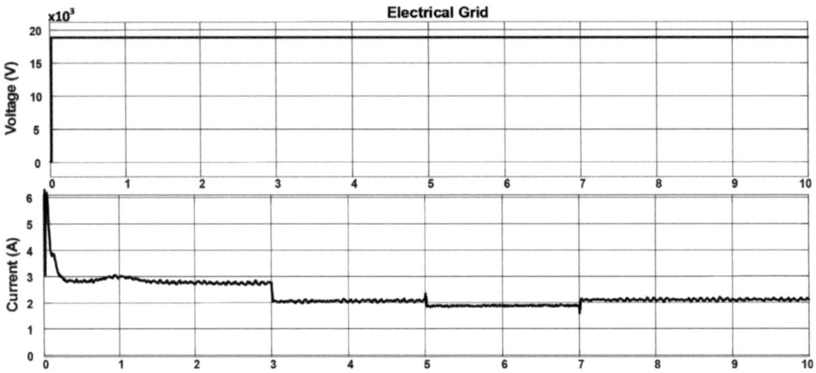

Figure 33.
RMS voltage and RMS current of electrical grid for case 6.

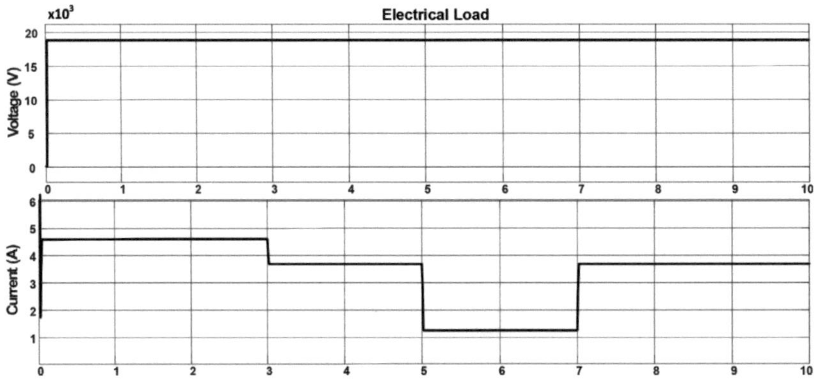

Figure 34.
RMS voltage and RMS current of electrical load for case 6.

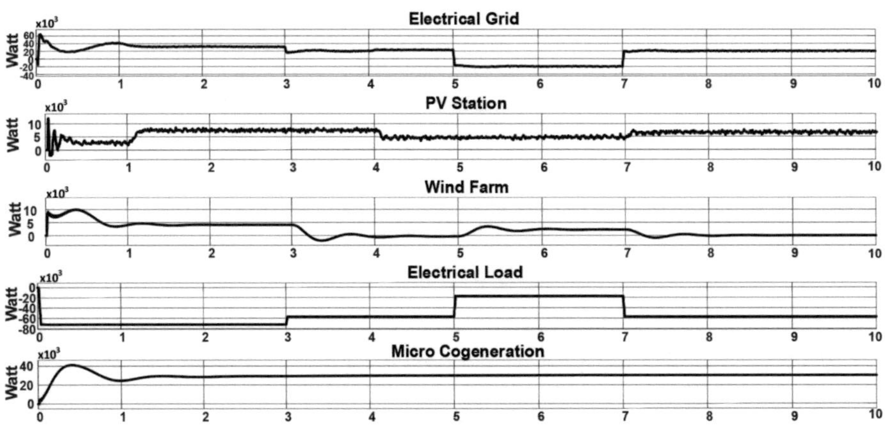

Figure 35.
The graph of active power of the elements in proposed system for case 7.

from **Figures 36–39**. The graph of wind speed, solar radiation, and temperature can be seen from **Figures 40** and **41**.

In this scenario wind speed, load demand, solar radiation, and temperature were variable. It is observed that despite the output power variations of the PV plant and WECS because of these variations, the system did not go to an unstable state.

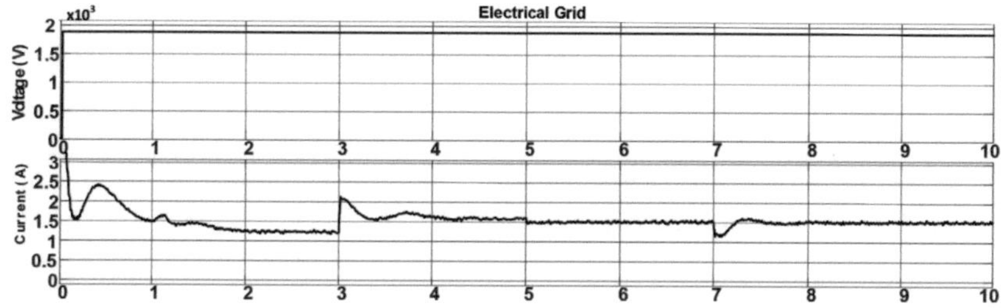

Figure 36.
RMS voltage and RMS current of electrical grid for case 7.

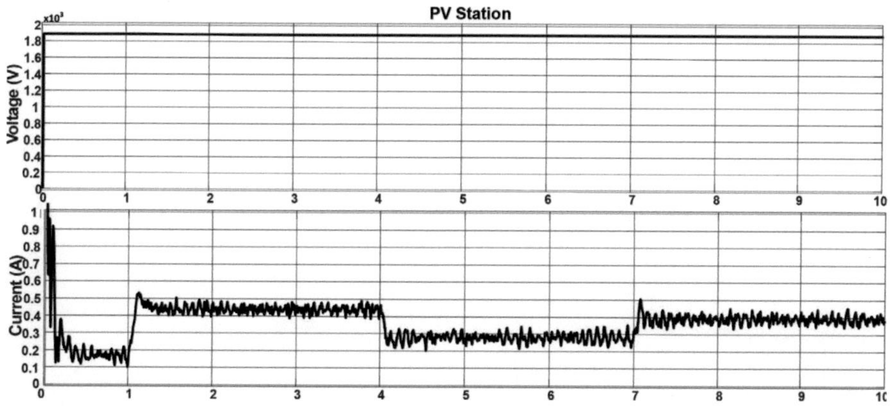

Figure 37.
RMS voltage and RMS current of PV station for case 7.

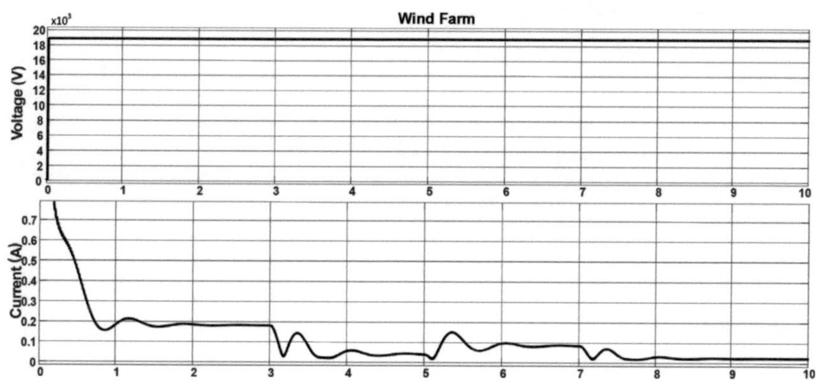

Figure 38.
RMS voltage and RMS current of wind farm for case 7.

Table 3 summarizes the power generation/consumption of the elements of the proposed system during simulations of the specified scenarios. Minus sign means power consumption; plus sign means power generation. From the simulation results, it is seen that the missing power due to the variations in wind speed and solar radiation is supplied from the electrical grid to the loads. It is observed that the proposed system that comprises a micro-cogeneration system is able to work stably despite all the fluctuations in the output powers of other sources such as WECS and PV plant and load demand.

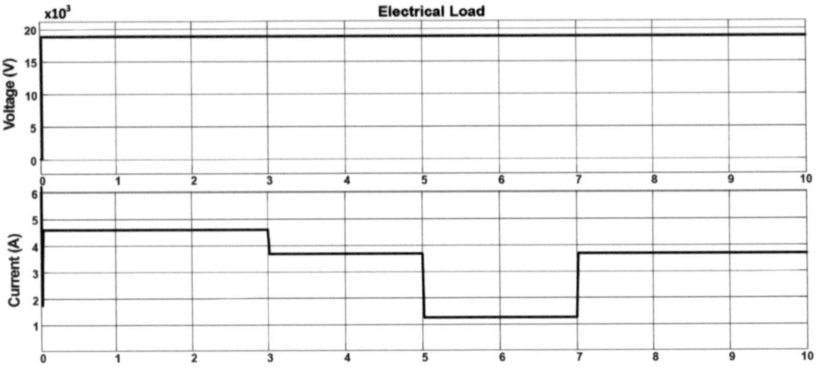

Figure 39.
RMS voltage and RMS current of electrical load for case 7.

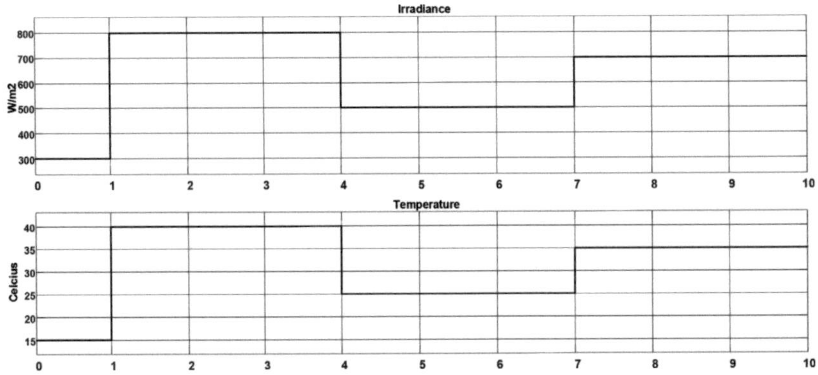

Figure 40.
Graph of solar radiation and temperature for case 7.

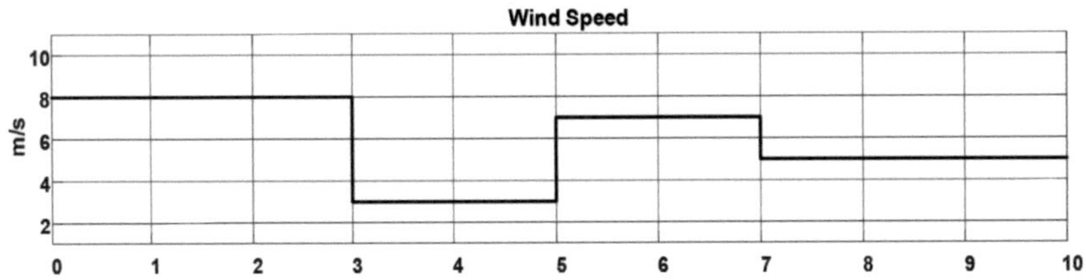

Figure 41.
Graph of wind speed for case 7.

Cases	PV system	WECS	Load	Micro-cogeneration system	Grid
Case 1	10 kW	10 kW	−50 kW	30 kW	—
Case 2	[1,10] kW	10 kW	−50 kW	30 kW	[−9,0] kW
Case 3	[1,10] kW	[0,10] kW	−50 kW	30 kW	[−19,0] kW
Case 4	[1,10] kW	10 kW	[−50,-5] kW	30 kW	[−40,0] kW
Case 5	10 kW	[0,10] kW	[−70,-45] kW	30 kW	[0,35] kW
Case 6	10 kW	10 kW	[−70,-20] kW	30 kW	[−30,20] kW
Case 7	[1,10] kW	[0,10] kW	[−70,-20] kW	30 kW	[−20,35] kW

Table 3.
Power generation and consumption values of the elements of the proposed system.

4. Conclusion

Renewable sources such as wind and solar have variable and intermittent characteristic by nature. For this reason, in microgrids supplied by these sources, there is a need for a compensating factor in order to ensure the stable operation of the microgrid for off-grid microgrids or the minimum energy demand from the electrical grid for on-grid microgrids. The expanding apprehension about climate change and energy subjection is driving specific policies to promote more efficient energy sources such as micro-cogeneration in many regions, particularly in the production of electricity. This paper proposes the use of micro-cogeneration system in a microgrid to support renewables in the microgrid.

The main contribution of this paper is that not only renewable energy sources such as solar and wind but also the micro-cogeneration system are used to supply electric energy. This study is carried out to analyze the operation of the micro gas turbine with solar and wind power generation systems. Firstly, a mathematical model of the micro gas turbine system, PV system, and WECS are given. After that, the control of the mentioned systems is explained. Lastly, the performance analysis of the proposed system is given via simulation results in seven different cases. The results show that the system is able to preserve the stable operation, while the parameters such as wind speed, solar radiation, temperature, and electrical load demand are changing.

Author details

Kemal Aygul[1], Burak Esenboga[2], Abdurrahman Yavuzdeger[3], Fırat Ekinci[3], Tugce Demirdelen[2]* and Mehmet Tumay[2]

1 Department of Electrical and Electronics Engineering, Cukurova University, Adana, Turkey

2 Department of Electrical and Electronics Engineering, Adana Alparslan Turkes Science and Technology University, Adana, Turkey

3 Department of Energy Systems Engineering, Adana Alparslan Turkes Science and Technology University, Adana, Turkey

*Address all correspondence to: tdemirdelen@atu.edu.tr

References

[1] Atmaca M. Efficiency analysis of combined cogeneration systems with steam and gas turbines. Energy Sources, Part A: Recovery, Utilization, and Environmental Effects. 2010;**33**(4): 360-369

[2] Sağlam G, Tutum CC, Kurtulan S. A new PI tuning method for an industrial process: A case study from a micro-cogeneration system. Energy Conversion and Management. 2013;**67**:226-239

[3] Lontsi F, Hamandjoda O, Fozao K, Stouffs P, Nganhou J. Dynamic simulation of a small modified Joule cycle reciprocating Ericsson engine for micro-cogeneration systems. Energy. 2013;**63**:309-316

[4] Capaldi P, Daliento A, Rizzo R. Prototype of a 20 kW cogenerator with reduced global cost. In: 2014 49th International Universities Power Engineering Conference (UPEC). IEEE. 2014. pp. 1-6

[5] Angrisani G, Marrasso E, Roselli C, Sasso M. A review on microcogeneration national testing procedures. Energy Procedia. 2014;**45**: 1372-1381

[6] Dang TT, Ruellan M, Prévond L, Ahmed HB, Multon B. Sizing optimization of tubular linear induction generator and its possible application in high acceleration free-piston stirling microcogeneration. IEEE Transactions on Industry Applications. 2015;**51**(5): 3716-3733

[7] Goyal R, Sharma D, Soni SL, Gupta PK, Johar D. An experimental investigation of CI engine operated micro-cogeneration system for power and space cooling. Energy Conversion and Management. 2015;**89**:63-70

[8] Angrisani G, Canelli M, Roselli C, Sasso M. Microcogeneration in buildings with low energy demand in load sharing application. Energy Conversion and Management. 2015;**100**:78-89

[9] Valenti G, Silva P, Fergnani N, Campanari S, Ravidà A, Di Marcoberardino G, et al. Experimental and numerical study of a micro-cogeneration Stirling unit under diverse conditions of the working fluid. Applied Energy. 2015;**160**:920-929

[10] Mustafa KF, Abdullah S, Abdullah MZ, Sopian K, Ismail AK. Experimental investigation of the performance of a liquid fuel-fired porous burner operating on kerosene-vegetable cooking oil (VCO) blends for micro-cogeneration of thermoelectric power. Renewable Energy. 2015;**74**:505-516

[11] Obi JB. State of art on ORC applications for waste heat recovery and micro-cogeneration for installations up to 100 kWe. Energy Procedia. 2015;**82**: 994-1001

[12] Angrisani G, Canelli M, Roselli C, Sasso M. Integration between electric vehicle charging and micro-cogeneration system. Energy Conversion and Management. 2015;**98**:115-126

[13] Darcovich K, Kenney B, MacNeil DD, Armstrong MM. Control strategies and cycling demands for Li-ion storage batteries in residential micro-cogeneration systems. Applied Energy. 2015;**141**:32-41

[14] Najafi B, Mamaghani AH, Baricci A, Rinaldi F, Casalegno A. Mathematical modelling and parametric study on a 30 kWel high temperature PEM fuel cell based residential micro cogeneration plant. International Journal of Hydrogen Energy. 2015;**40**(3):1569-1583

[15] Özgirgin E, Devrim Y, Albostan A. Modeling and simulation of a hybrid photovoltaic (PV) module-electrolyzer-PEM

fuel cell system for micro-cogeneration applications. International Journal of Hydrogen Energy. 2015;**40**(44): 15336-15342

[16] Najafi B, Mamaghani AH, Rinaldi F, Casalegno A. Fuel partialization and power/heat shifting strategies applied to a 30 kWel high temperature PEM fuel cell based residential micro cogeneration plant. International Journal of Hydrogen Energy. 2015; **40**(41):14224-14234

[17] Bouvier J-L, Michaux G, Salagnac P, Nepveu F, Rochier D, Kientz T. Experimental characterisation of a solar parabolic trough collector used in a micro-CHP (micro-cogeneration) system with direct steam generation. Energy. 2015;**83**:474-485

[18] Prinsloo G, Dobson R, Mammoli A. Model based design of a novel Stirling solar micro-cogeneration system with performance and fuel transition analysis for rural African village locations. Solar Energy. 2016;**133**:315-330

[19] Di Marcoberardino G, Roses L, Manzolini G. Technical assessment of a micro-cogeneration system based on polymer electrolyte membrane fuel cell and fluidized bed autothermal reformer. Applied Energy. 2016;**162**:231-244

[20] Zheng CY, Wu JY, Zhai XQ, Yang G, Wang RZ. Experimental and modeling investigation of an ICE (internal combustion engine) based micro-cogeneration device considering overheat protection controls. Energy. 2016;**101**:447-461

[21] Ferreira AC, Nunes ML, Teixeira JCF, Martins LASB, Teixeira SFCF. Thermodynamic and economic optimization of a solar-powered Stirling engine for micro-cogeneration purposes. Energy. 2016;**111**:1-17

[22] Bilgen S, Sarıkaya İ. The use and its impact on the environment of cogeneration as an important element for a clean and sustainable energy future. Energy Sources, Part A: Recovery, Utilization, and Environmental Effects. 2017;**39**(21): 2078-2086

[23] Yu T, Tong J-P. Auto disturbance rejection control of microturbine system. In: Power and Energy Society General Meeting-Conversion and Delivery of Electrical Energy in the 21st Century, 2008 IEEE. 2008. pp. 1-6

[24] Li JJ. Modeling and simulation of micro gas turbine generation system for grid connected operation. In: Power and Energy Engineering Conference (APPEEC), 2010 Asia-Pacific. IEEE. 2010. pp. 1-4

[25] Gaonkar DN, Patel RN. Modeling and simulation of microturbine based distributed generation system. In: 2006 IEEE Power India Conference. 2006. p. 5

[26] Bellia H, Youcef R, Fatima M. A detailed modeling of photovoltaic module using MATLAB. NRIAG Journal of Astronomy and Geophysics. 2014; **3**(1):53-61

[27] Singh G, Siwasia M, Vyas S, Kumar R. Implementation of a fuzzy logic based perturb and observe approach for maximum power point tracking of solar PV modules. In: 2016 IEEE 7th Power India International Conference (PIICON); IEEE. 2016. pp. 1-6

[28] Nedumgatt JJ, Jayakrishnan KB, Umashankar S, Vijayakumar D, Kothari DP. Perturb and observe MPPT algorithm for solar PV systems-modeling and simulation. In: 2011 Annual IEEE India Conference (INDICON); IEEE. 2011. pp. 1-6

[29] Hur S. Modelling and control of a wind turbine and farm. Energy. 2018; **156**:360-370

[30] Mosa MA, Elsyed AA, Amin AM, Ghany AA. Variable speed wind turbine pitch angle controller with rate limiter anti-windup. In: 2016 Eighteenth International Middle East Power Systems Conference (MEPCON); IEEE. 2016. pp. 95-100

[31] Dhar MK, Thasfiquzzaman M, Dhar RK, Ahmed MT, Al Mohsin A. Study on pitch angle control of a variable speed wind turbine using different control strategies. In: 2017 International Conference on Power, Control, Signals and Instrumentation Engineering (ICPCSI); IEEE. 2017. pp. 285-290

Chapter 6

Hybrid Maritime Microgrids: A Quest for Future Onboard Integrated Marine Power Systems

Thomas Caravella, Christopher Austell, Christian Brady-Alvarez and Salem Elsaiah

Abstract

The following is a comprehensive analysis which details potential ways for the maritime industry to begin to phase out AC power generation and distribution on new vessels over a short period of time. Therefore, the vessels of the future should consider transitioning into DC power generation and distribution. During the transition from AC shipboard systems to DC shipboard systems, there will be a time during which the vessels will be run by "hybrid" shipboard power systems, which utilize a mixture of AC and DC power. These systems are known as integrated marine power systems (IMPS) or hybrid maritime microgrid architectures, since they represent a distribution system or a part thereof. This study presents a state of the art of maritime systems, emphasizing on the design aspects of hybrid maritime microgrids, summarizing the advantages, disadvantages, and the challenges that planners may face when it comes to the vessels of the future. This study also reviews remedies that have been recently proposed in the literature to overcome such challenges. In addition, this work reports on the problem of service restoration of shipboard power systems and introduces directions on how to enhance the survivability of maritime power systems using techniques based on distribution system reconfiguration.

Keywords: integrated marine power systems, hybrid microgrids, maritime microgrids, distributed generators, renewable energy resources, power electronics, energy systems

1. Introduction

The dynamic evolution of the electrical loads aboard ships and vessels within the marine industry in the twenty-first century calls for further research and development of integrated marine power systems, with DC distribution and electrical propulsion components. The emerged IMPS comprise of the conventional AC generators and propulsion systems, in addition to DC distribution systems along with DC loads such as switchboards. The strong consideration of DC distribution systems in place of the commonly used AC within marine power systems has gained notice due to the much advancement in today's power electronic circuitry.

The demands for sufficient electrical power in future ship designs varying from warships and naval aircraft carriers to oil tankers and transport vessels are of main importance to current ship designers. From AC generators to the propellers, today's loads predominantly exist within the DC section of the system. With the introduction of advanced communication systems, electromagnetic aircraft launch systems, and electrical weaponry onto naval vessels, high-reliability integrated power systems that can meet varying electrical load demands are highly sought. Power system architectures are dependent upon load types, and the viability of DC distribution, such as medium- or low-voltage DC power systems, in meeting upcoming shipboard electrical demand forecasts is of focus. The paradigm shift of traditional AC system constructions to DC entails both operational benefits and, on the other hand, any accompanying risks. One form of system design does not by any means entitle an industry standard, as vessel functions vary from ship to ship and boat to boat. Ideal system configurations coincide with the shipboard loads and are characterized by high reliability, ease of maintenance, both fuel and payload efficiency, and lower costs and low, or optimally absent, emissions.

In [1], ABB has reported that an Onboard DC Grid can allow for fault current clearance within a time window of a maximum of 40 ms, as compared to 1 second for traditional AC circuits. Furthermore, because the speed of prime movers must be locked at approximately 60 Hz in AC marine systems, the lowest fuel consumption will be achieved operation about 85% of the rated load. Because DC buses are at no frequency, the prime mover speed can be adjusted according to demand, all the while correspondingly adjusting the generator excitation current in the same respects. Given this commodity of optimization speed, the operating window of the prime movers can be brought down to about 50% of the rated load with no increase in fuel consumption [2].

In recent years, there has been an increasing interest in the design and analysis of microgrids in general [3–5], with particular interest on the implementation of hybrid microgrids (*AC and DC microgrids*) within the maritime industry [6]. For instance, the Italian Navy, by the use of a system nicknamed naval package (NP), has successfully corroborated simulation results with experimental data in determining the validity of a medium-voltage DC shipboard IMPS with the use of dual three-phase 2.15 MVA generators. The results showed that the generator produced quality DC outputs with the introduction of faults and specified rectifier arrangements reduced performance degradation and increased system fault tolerance [6].

Several operations are constantly performed on shipboard power systems. Of these, service restoration and reliability improvement may be of most concern. Numerous marine power distribution systems are characterized by radial or weakly meshed topological structure. The unidirectional power flow in the radial distribution systems facilitates the coordination of the protective devices used at the distribution system level. Nevertheless, this radial topological structure makes distribution systems less reliable, compared to transmission systems, which are highly interconnected systems. Despite the unidirectional power flow in the radial distribution systems, the failure of any single component between the load point and the source node may cause service interruptions, which could lead to disconnection of several load points. Distribution systems are constantly equipped with two types of switches: sectionalizing switches and tie switches. The sectionalizing switches are normally closed; however, the tie switches are normally open but can be closed to routing the power and meeting the power demand during abnormal conditions. Distribution system reconfiguration can be used to minimize the duration and frequency of service interruptions and thereby enhance the reliability of the distribution system and the quality of service. By distribution system reconfiguration, we denote the process of changing the topology of the distribution network by altering the status of sectionalizing and tie switches to achieve certain objectives [4]. Of these objectives, reliability improvement, service restoration, and survivability enhancement are of great concern. Generally speaking, reconfiguration and service restoration of marine power systems are carried out in a similar manner to that used for terrestrial power systems, with few restrictions and constraints.

In [7], a method for load sharing in hybrid microgrid systems using control loops is presented. A method for maritime microgrid reconfiguration is presented in [8] using genetic algorithms and heuristic techniques. The work presented in [8] is implemented on a small shipboard power system, and a modified CERTS microgrid system including distributed generators has been used. Methods for optimal feeder reconfiguration for terrestrial distribution systems using intelligent and heuristic methods are also proposed in [3–5]. In [9], a method for optimal sizing of renewable energy resources for next-generation seaports is presented. The renewable energy resources used in [9] are modeled using HOMER platform. A two-stage technique for isolated microgrid systems is introduced in [10]. The method proposed in [10] used graph theory and binary firefly algorithm to perform minimum load curtailment in the isolated microgrid. Methods for service restoration of marine power systems have also been presented in [11–13].

This study is organized as follows. Section 2 introduces a state of the art of modern power systems. Section 3 presents, discusses, and lists the unique features of next-generation marine power systems. Section 4 introduces the concept of maritime microgrids and discusses briefly the composition of such maritime microgrids. It also highlights the main differences between the conventional shipboard systems and next-generation maritime microgrid systems by giving examples and illustrations and single-line diagram of a maritime microgrid system. Section 5 presents modeling aspects for some of the marine power system components. Section 6 highlights feature studies for marine power systems followed by conclusions.

2. Modern marine power systems: state of the art

Modern ships consume a substantial amount of power due to a wide range of equipment needed to operate each vessel. Some of these equipment include cargo equipment (*such as cargo pumps and cargo cranes*) as well as other necessities including electric heaters, control equipment, and motors for pumps, compressors, or propulsion. Modern vessels always have a main switchboard and an auxiliary switchboard (*emergency switchboard*), which are connected via numerous circuit breakers (CB as in **Figure 1**). Usually the switchboards receive power from large AC synchronous generators. Both the main and emergency switchboards are interlocked with an automatic bus tie breaker that disconnects the two should power loss has been sensed on the main switchboard. The main generators are all connected in parallel to the main switchboard as depicted in **Figure 1**; however, only the emergency generator, not shown in **Figure 1**, is connected to the emergency switchboard. The emergency generator is periodically run in order to conduct weekly tests as well as monthly load tests; otherwise, the emergency generator is only powered up in the event of a power loss emergency. The main generators are connected to the switchboard one-by-one using circuit breakers as shown in **Figure 1** until there are enough online to power all electrical loads of the vessel. From the switchboards, power is then distributed to all appliances. There are multiple transformers on vessels used to step down the voltage for equipment, which typically require less voltage. Much like landside power plants, automatic voltage regulators (AVR) and governors are implemented in order to control the generator's power, frequency, and voltage output. On most vessels, the only DC power is more commonly used to power computerized control equipment as well as the DC excitation of the synchronous generators, the general alarm, emergency lighting, and the electronic integrated circuits. Each of these systems use minimal amounts of power, which highlights the notion that modern marine power systems are almost entirely supplied by AC power. From the emergency switchboard, power flows through a rectifier which converts it from AC to DC, which trickle charges the batteries, such as battery bank (BT1 in **Figure 1**), that power the systems previously described.

One issue commonly found in AC distribution systems is the power loss that is almost inevitable due to the nature of the load and therefore the power factor. It is noteworthy to mention here that numerous vessels' power systems have extremely poor power factors, which should be corrected in order to increase efficiency. This can be accomplished using capacitor banks or synchronous condensers, which will bring a severely lagging power factor closer to unity. Unfortunately, such power factor correction devices will likely still waste resources, such as fuel, power, and money, which are never desirable. Section 5 briefly discusses modeling aspects of capacitor banks for power factor improvement of marine power systems.

As far as for DC systems, it is appropriate to highlight here that DC power is often inefficiently transmitted and distributed over large distances, which is one of the main reasons why landside AC power transmission is used. Fortunately, for

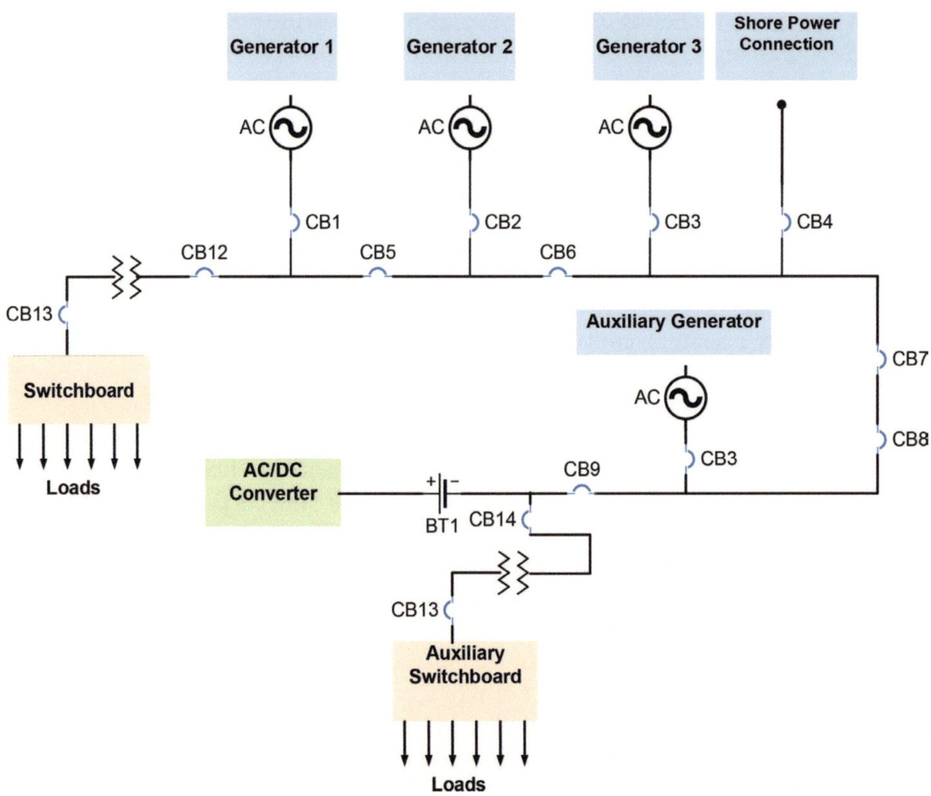

Figure 1.
Single-line diagram of a modern marine power system.

shipboard electric power systems, DC power distribution is a highly viable option. Unlike landside power generation and distribution, vessel power grids do not travel across massive distances. For example, one of the longest vessels ever made is only 1500 feet long, which is miniscule compared to the massive distances, or hundreds of miles, over which landside power companies are expected to transmit and distribute electric power.

3. Future marine power systems

Due to the added regulations regarding the increase in efficiency and decrease in pollution of all vessels, it has become evident why most companies desire and are actively searching for better shipboard microgrid architectures. The overall goal of future marine power systems would be to establish and use an almost completely DC marine power system. This is due to the numerous potential benefits it would allow in both efficiency and pollution if properly designed [14–19]. Numerous vessels now deploy diesel engine propulsion; however, they can likely be made more efficient in the future if they were to be switched to electric propulsion under the umbrella of all electric ships (AES). Electric propulsion may improve fuel consumption as well as the dynamic performance of the ship [14]. Since ships that use electric propulsion consume most of the power generated for propulsion, it is imperative to ensure that the shipboard microgrid of such a vessel is capable of handling the load dynamics, which are connected to electric propulsion as well as the immense amount of power needed to propel a vessel utilizing electric propulsion technologies.

With the use of energy storage systems (ESS), such as battery banks and fuel cells, synchronous generators can be operated at optimal speeds, which allow for improved efficiency and reduced pollution, as a result, allowing a greater reduction in distribution losses as well as increased reliability [7]. The use of energy storage systems is beneficial as they allow for continuous power flow to the connected loads in the event of a power generator failure.

Transformers, which are AC/AC converters, are used in conventional shipboard AC power systems. Such devices would be replaced with DC/DC power converters for the shipboard DC distribution systems of the future, which would allow the voltage to be stepped down (buck converters, for instance) or stepped up (boost converters, for instance), to the required levels. The voltage levels may also be stepped up and stepped down to achieve certain requirements by the means of buck/boost or cuck converters. On the other hand and for specific AC loads, DC/AC inverters would be required. Benefits of using such converters may include increased power compensation and better frequency regulation.

It is appropriate to mention that the modern and future maritime systems have many similarities in structure; however, they also possess few differences. For instance, as shown in **Figure 2**, one of the major noticeable differences is that in future marine power systems, both the main and the auxiliary or emergency switchboards receive and distribute DC power as opposed to the AC power utilized on the modern marine power system example depicted in **Figure 1**. Another key difference is the use of an energy storage system, which powers the switchboard in the next-generation marine power systems as can be seen from **Figure 2**. Though only few ships are currently deploying electric propulsion, the ships of the future should solely use electric propulsion due to its higher efficiency. Some advantages of using DC-powered systems instead of the conventionally used AC systems on next-generation shipboard systems are summarized in **Table 1** [1, 2, 19, 20].

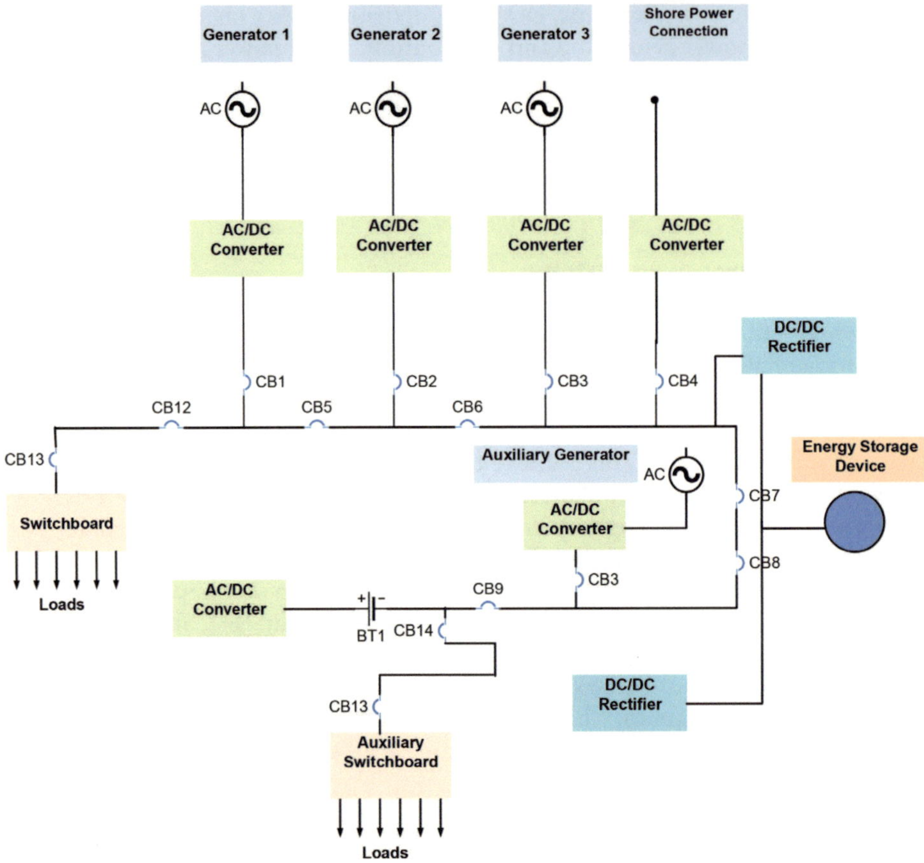

Figure 2.
Single-line diagram of a next-generation marine power system.

Item	Advantages of using DC systems vs. AC systems
1	Eradication of frequency-related issues and synchronization of sources
2	Proper control of drive systems over a wide range of speeds due to the advancement in power electronic devices
3	Potential reduction in overall size and rating of switchgear
4	Generators operate at or near unity power factor; and thereby reactive power compensation may be controlled properly
5	Overall reduced system size in general, allowing for larger cargo space
6	Virtual inertia of power electronic devices may be deployed to enhance the overall system stability
7	Replacement of large distribution transformers with small-sized power electronic converters

Table 1.
Advantages of using DC-powered marine systems.

4. Maritime microgrids

In general, DC microgrids characteristically are self-sustaining given the ability to work given a grid-connected or islanded-mode condition. The feature of these self-thriving distribution systems emanates from the connected renewable energy resources and energy storage devices. The general architecture of the emerged maritime microgrids is depicted in **Figure 3**. As shown in **Figure 3**, most maritime microgrids consist of conventional generators, several power electronic circuitries, and numerous AC and DC loads such as the propellers and propulsion system as a whole.

In **Figure 3**, the maritime microgrid system under study consists of two synchronous generators operating in parallel and connected via a sectionalizing switch. The switch is normally open and can be closed during harsh conditions, as in case of breaking, for example. The hybrid maritime microgrid system shown in **Figure 1** is driven by power electronic circuitry, which is mainly consisting of the following:

- AC/DC full-bridge inverter system.
- Energy storage system EES, and in our case we assumed a battery system.
- DC/DC boost converter(s).
- DC/AC inverter(s). One of the main functions of the DC/AC inversion systems is to feed the propellers.

A lot of research is being conducted to optimize the operation of maritime microgrids such as enhancing reliability and survivability, improving stability using virtual synchronous generators, and minimizing power curtailment. References [1–13, 20] provide literature on some of these techniques.

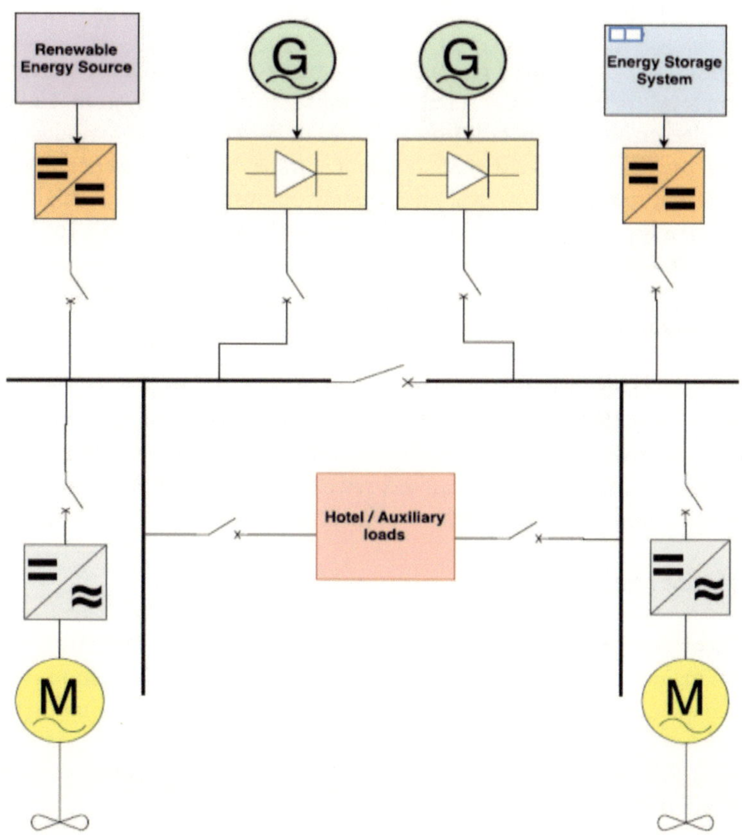

Figure 3.
Hybrid maritime microgrid system.

5. Review of marine power system components

In the subsequent section, modeling aspects of some components used on most of the current and future marine power systems are reviewed and briefly discussed.

5.1 Load model

The active and reactive power loads on distribution networks can be represented as constant power, constant current, constant impedance, or a mixture of these types. Hence, the load model in distribution systems can be generally represented as by an exponential form as

$$P_k = P_{ref}\left(\frac{V_k}{V_{ref}}\right)^{\alpha} \quad (1)$$

$$Q_k = Q_{ref}\left(\frac{V_k}{V_{ref}}\right)^{\beta} \quad (2)$$

where V_{ref} is the reference bus voltage; V_k is the operating voltage at bus k; P_{ref} and Q_{ref}, respectively, are the active and reactive power consumptions at the reference bus; and α and β are exponents on which the load characteristic can be determined. That is, constant power load model can be found by setting α and β equal to 0. Further, constant current model is obtained by setting α and β equal to 1. Moreover, constant impedance load model is obtained by setting α and β equal to 2.

5.2 Cogeneration model

Cogenerators or distributed generators are small-scale sources of energy used at distribution level to increase the reliability and security of the system. Cogenerators can be modeled either as constant power nodes or constant voltage nodes. However, according to the IEEE standard 3636, cogenerators are modeled as constant power loads. Therefore, in several studies, particularly planning studies, it might be appropriate to adopt modeling of cogeneration such as photovoltaic, for instance, as constant power loads.

5.3 Static VAR model

As was mentioned earlier in Section 2, capacitor banks and other types of static VAR compensators (SVC) are utilized on maritime distribution systems to boost bus voltages and keep reactive power limits in the desired range. Capacitor banks can be modeled as constant capacitance devices and may be represented by current injections to the node to which they are connected.

5.4 Transformer model

Transformers are modeled using their per unit synchronous reactance. Using per unit system makes including transformers in power system analysis much easier as the problem of transferring reactances on both sides of the transformer is eliminated.

6. Future studies

Based on our discussion so far, we have concluded that maritime power systems are somewhat different from their typical terrestrial counterparts. In particular,

Item	Challenges
1	The penetration of intermittent renewable energy resources and storage devices has increased in nowadays shipboard systems. Consequently, accurate models need to be developed to accommodate such an increase
2	To ensure reliable operation of next-generation shipboard systems, numerous studies should be carried out in a real-time frame using high-computation facilities and parallel computing
3	More innovative techniques for optimal load flows, service restoration, and reconfiguration need to be developed. This requires the development of more testbeds, particularly for large-scale realistic marine power systems
4	Frequency-related issues of power electronic devices are another area of research for future shipboard systems. Effective inverter topologies that minimize stress ratios and switching losses constitute another research path for future maritime systems
5	For stabilize operation of the hybrid power systems, more work needs to be carried out in the area of power system stability using the concept of virtual inertia. Proper control design and innovative communication protocols would help utilizing the concept of virtual synchronous generators to maintain the stability limits of a given marine electric system

Table 2.
Examples and challenges of future research in marine systems.

generally speaking, maritime systems consist of generation and distribution, but no transmission, which is very distinct from the territorial electric power systems. The generation and distribution are coupled with distribution feeders in order to transfer the power from the synchronous generators to the loads. Hybrid maritime distribution systems (*hybrid maritime microgrids*) consist also of several power electronic devices such as converters, inverters, rectifiers, and switchboards. Instrumentation and control circuits are used on ships to ensure safe operation of the ship. A typical maritime power system is in fact an isolated power system, which is largely different from the terrestrial power systems. Therefore, survivability of such a system during abnormal circumstances is of great concern for marine power system planners. Enhancing reliability and survivability through feeder reconfiguring of marine distribution systems, however, needs flexible and reliable techniques to accommodate all of the aforementioned characteristics of modern marine power systems. In order to perform studies on any marine power system, a power flow solution is constantly required. Nonlinear power flow and DC power flows are commonly used in the literature to carry out optimization studies on terrestrial and shipboard systems. In recent literature, enhanced linearized power flows [21, 22] have also been used as they compensate for most of the drawbacks of the conventional DC power flows. A lot of research needs to be performed in order to secure transition from AC marine power systems to hybrid and then DC-powered marine power systems. This research may include developing methods for enhancing reliability and survivability, improving stability using virtual synchronous generators, and minimizing power curtailment on ships using real-time data and testbeds. **Table 2** summarizes some challenges and also lists examples of research that need to be carried out in order to accommodate the emerged maritime power systems.

7. Conclusion

This study has detailed some potential ways for the maritime industry to begin to phase out AC power generation and distribution on new vessels over a short period of time. Over short distances, DC power provides less power loss, no harmonics, and more control over equipment, which are some of the major reasons why many industry officials are conducting vast research in this specific area. Therefore, the vessels of the future should consider transitioning into DC power generation and distribution. Nonetheless, during the transition from an AC shipboard power system to a DC shipboard power system, there will be a time during which the vessels will be run by a hybrid shipboard power system, which utilizes a mixture of both AC and DC power. These hybrid systems are known as integrated marine power systems or hybrid maritime microgrids.

Some advantages of hybrid maritime microgrids include higher efficiency, optimum fuel consumption, improved reliability, and more importantly improved survivability. This book chapter provides a quest for future maritime microgrids and integrated marine power systems. Though the power system literature is rich for terrestrial microgrids, it has fallen behind in providing a solid background for the future maritime systems. One of the main objectives of this review study is to pave the way for the researchers by supplying them with recent technologies, visions, and applications for future maritime microgrids. This study has presented a state of the art for maritime microgrids, emphasizing on the design aspects of hybrid maritime microgrids and summarizing the advantages, disadvantages, and the challenges that planners face when integrating renewable energy resources into existing marine power systems. Moreover, this review study has paved the way for mariners and researchers by supplying them with recent technologies, visions, and applications for future maritime integrated power systems and maritime microgrids. This work has also presented and discussed issues associated with the design and control of future maritime microgrids as envisioned by the US Navy near-term development plan of 2025 and the long-term plan of 2035. In addition, this study presented some of the challenges that both current and future IMPS are facing and reviewed some of the remedies that have been recently proposed in the literature to overcome such challenges. This study has also reported on the problem of feeder reconfiguration and service restoration of shipboard power systems and introduces directions on how to enhance the reliability and survivability of maritime power systems using distribution system reconfiguration.

Acknowledgements

This project represents a part of an ongoing research at the undergraduate level at the State University of New York, Maritime College. The corresponding author is grateful to Cadets Thomas Caravella and Christopher Austell for the time and effort they have dedicated to work on this project beside their regimental obligations. The corresponding author is also grateful for Dr. Taha Selim Ustun for the kind invitation to contributing to this book. His insightful comments and timely feedback of the revised chapter have revamped the quality of this work.

Author details

Thomas Caravella[1], Christopher Austell[1], Christian Brady-Alvarez[2] and Salem Elsaiah[1*]

1 Department of Engineering, Maritime College (SUNY Maritime), State University of New York, NY, United States

2 Northrop Grumman, Melbourne, FL, United States

*Address all correspondence to: selsaiah@sunymaritime.edu

References

[1] ABB. The step forward: Onboard DC Grid. Report. 2011

[2] Prenc R, Cuculic A, Baumgartner I. Advantages of using a DC power system on board ship. Pomorski zbornik. 2016;**52**:83-97

[3] Elsaiah S, Mitra J. A method for minimum loss reconfiguration of radial distribution systems. In: Proceedings of the IEEE Power and Energy Society General Meeting and Exposition, Denver, Colorado, USA. 2015

[4] Elsaiah S, Benidris M, Mitra J. Reliability improvement of power distribution system through feeder reconfiguration. In: Proceedings of the 13th IEEE International Conference on Probabilistic Methods Applied to Power Systems, PMAPS. Durham, UK; 2014

[5] Elsaiah S, Benidris M, Mitra J. A method for reliability improvement of microgrids. In: Proceedings of the IEEE 19th Power System Computation Conference. Genoa, Italy; 2016

[6] Sulligoi G, Tessarolo A, Benucci V, Barret M, Rebora A, Taffone A. Modeling, simulation and experimental validation of a generation system for medium-voltage DC integrated power systems. IEEE Transactions on Industry Applications. 2010;**46**:1304-1310

[7] Wei J, Yu Z. Load sharing techniques in hybrid power systems for DC microgrids. In: Proceedings of the Asia-Pacific Power and Energy Engineering Conference, IEEE. 2011. pp. 1-4

[8] Shariatzadeh F, Zamora R, Srivastava AK. Real-time implementation of microgrid reconfiguration. In: Proceedings of the North American Power Symposium, IEEE. 2011. pp. 1-6

[9] Ahamad NB, Othman M, Vasquez JC, Guerrero JM, Su C-L. Optimal sizing and performance evaluation of renewable energy based microgrid for future seaports. In: Proceedings of International Conference on Industrial Technology (ICIT), IEEE. 2018. pp. 1-6

[10] Hari Kumar R, Ushakumari S. A two stage algorithm for optimal management of isolated microgrid. In: Proceedings of International Conference on Signal Processing, Informatics, Communication and Energy, IEEE. 2015. pp. 1-5

[11] Su C-L, Lan C-K, Chou T-C, Chen C-J. Design of a multi-agent system for shipboard power systems restoration. In: Proceedings of the 50th Industrial and Commercial Power Systems Technical Conference, IEEE/IAS. 2014

[12] Jiang Y, Jiang J, Zhang Y. A novel fuzzy multiobjective model using adaptive genetic algorithm based on cloud theory for service restoration of shipboard power systems. IEEE Transactions on Power Systems. 2012;**27**(2):612-620

[13] Butler-Purry KL, Sarma NDR. Self-healing reconfiguration for restoration of naval shipboard power systems. IEEE Transactions on Power Systems. 2004;**19**(2):754-762

[14] Othman M, Anvari-Moghaddam A, Guerrero J. Hybrid shipboard microgrids: System architectures and energy management aspects. In: Proceedings of the IEEE 43rd Annual Conference of the IEEE Industrial Electronics Society, (IECON), IEEE. 2017. pp. 1-6

[15] Grigoryev AV, Malyshev SM, Zaynullin RR. Unified ship power grids with alternators and DC power distribution. In: Proceedings of the IEEE International Conference on Industrial Engineering, Applications and Manufacturing (ICIEAM), IEEE. 2017. pp. 1-3

[16] Guerrero JM, Jin Z, Liu W, Othman MB, Savaghebi M, Anvari-Moghaddam A, et al. Shipboard microgrids: Maritime islanded power systems technologies. In: PCIM Asia, 2016. PCIM Asia 2016, 28-30 June 2016, Shanghai, China. 2016. pp. 1-8

[17] Jin Z, Savaghebi M, Vasquez JC, Meng L, Guerrero JM. Maritime DC microgrids: A combination of microgrid technologies and maritime Onboard power system for future ships. In: Proceedings of the IEEE 8th International Power Electronics and Motion Control Conference (IPEMC-ECCE Asia), IEEE. 2016. pp. 1-6

[18] Shekhar A, Ramírez-Elizondo L, Bauer P. DC Microgrid Islands on ships. In: Proceedings of the IEEE Second International Conference on DC Microgrids (ICDCM); 2017

[19] Brady-Alvarez C, Elsaiah S. A review of renewable energy resources and their applications. In: Scundurra G, editor. Focus on Renewable Energy Sources. New York: Nova Science Publishers Inc.; 2018. pp. 131-156. ISBN: 978-1-53613-802-3

[20] Jin Z, Sulligoi G, Cuzner R, Meng L, Vasquez JC, Guerrero JM. Next-generation ship-board DC power system. In: IEEE Electrification Magazine. 2016. pp. 45-57

[21] Elsaiah S, Cai N, Benidris M, Mitra J. A fast economic power dispatch method for power system planning studies. IET Generation Transmission and Distribution. 2015;**9**(5):417-426

[22] Elsaiah S, Benidris M, Mitra J. An analytical approach for placement and sizing of distributed generators on power distribution system. IET Generation Transmission and Distribution. 2014;**8**(6):1039-1049

Section 3

Electric Vehicles

Chapter 7

Vehicle Electrification: Technologies, Challenges, and a Global Perspective for Smart Grids

Vitor Monteiro, Jose A. Afonso, Tiago J.C. Sousa, Luiz L. Cardoso, Jose Gabriel Pinto and Joao L. Afonso

Abstract

Nowadays, due to economic and climate concerns, the private transportation sector is shifting for the vehicle electrification. For this new reality, new challenges about operation modes are emerging, demanding a cooperative and dynamic operation with the power grid, guaranteeing a stable integration without omitting the power quality. Besides, new attractive and complementary technologies are offered by the vehicle electrification in the context of smart grids, valid for both on board and off board systems. In this perspective, this book chapter presents a global perspective and deals with challenges for the vehicle electrification, covering the key technologies toward a sustainable future. Among others, the flowing topics are covered: (1) Overview of battery charging systems, including on board and off board systems; (2) State of the art of communication technologies for application in the context of vehicular electrification, smart grids and smart homes; (3) Challenges and opportunities concerning wireless power transfer with bidirectional interface to the electrical grid; (4) Future perspectives about bidirectional power transfer between electric vehicles (vehicle to vehicle operation mode); (5) Unified technologies, allowing to combine functionalities of a bidirectional interface with the electrical grid and motor driver based on a single system; and (6) Smart grids and smart homes scenarios and accessible opportunities about operation modes.

Keywords: vehicle electrification, smart grids, smart homes, communication technologies, wireless power transfer, renewable energy sources, power quality, power electronics, energy storage systems

1. Introduction

Nowadays, the transport sector is responsible by 33% of final energy consumption in the 28 countries of the European Union (EU28), where road transports represent about 82%, contributing to about 27% of the total final energy consumed in EU28 [1]. Associated with this consumption is the emission of greenhouse gases for the atmosphere, contributing for the global warming, as well as for deteriorate living conditions on the planet. Indeed, the environmental problems are affecting the societies around the world, obliging to change the paradigm targeting moderating the greenhouse gas emissions [2, 3] Globally, the transports sector contributes in 26% for the final energy consumption and 13.1% for the total CO_2 emissions [4], and in particular, contributed nearly 21% of the EU total emissions of CO_2 [5]. As a contribution to overcome this paradigm, the vehicle electrification (electric and hybrid electric vehicles) is pointed-out by many specialists as a prominent solution to reduce the CO_2 emissions [6–8] and to support the future transportation sector [9–11]. From the different solutions offered by the vehicle electrification, plug-in electric vehicles (EV) and plug-in hybrid EV are particularly interesting due to the capacity to be charged directly from the electrical grid. In fact, the changing of paradigm for the electric mobility is already underway and the global sales of EV reached the 174,000 units in the first trimester of 2018 with a consistent growing of about 67% by year. In terms of geographic distribution, China leads the sales of EV followed by United States and Japan, and by Norway and Germany in the EU [12]. The global sales of plug-in EVs reached 2.1 million units in 2018, growing 64% in relation to 2017 [13]. At the end of 2018, the global fleet of plug-in EVs reaches the mark of 5.4 million [13].

Also as a complement to revolutionize this paradigm change, new technologies are appearing targeting a common harmonious objective for the smart grids [14]: decrease of greenhouse gas emissions. Thereby, besides electric mobility, other technologies are influent for the same purpose, as renewable energy sources (RES) and energy storage systems (ESS), at residential or industrial level, strategic installed as support of the electrical power grid [15–19] These three technologies are accepted as elementary pillars for a profound and exciting revolution of paradigm toward, each more, smart grids as well as smart homes, where the developments in the industrial and electronics field are indispensable [20–23]. Additionally, based on the presence of vehicle electrification, RES, and ESS when engaged with smart grids and smart homes, further advances in the field of information and communication technologies are welcome, where the Internet of Things (IoT) concept appears as a key contribution to help to obtain an autonomous, dynamic, and flexible electrical grid [24–27].

Analyzing the introduction and maturation of RES along the last decades, especially the imposition offered by wind and solar, the power production from these sources has grown expressively, signifying a vital input for enhancing the power management and the energy necessities, both at residential and industrial level. In an ample perception, it is clear that the involvement of RES represents a positive effect for disseminating the new paradigm of smart grids [28]. This is also valid from the point of view of a microgrid, where the planned distribution of RES also represents a pertinent participation for an optimal power control process in smart grids [29]. Nonetheless, RES has a disadvantage that cannot be solved by itself: the dependence on weather conditions that is reflected in the intermittent power production. In this way, the inclusion of ESS technologies is indispensable to establish a support between the power production and consumption in collaboration with the necessities of the electrical grid (more precisely, for instance, the necessities of smart homes). In this context of using ESS to balance the power production and consumption, it is established an efficient engagement of unpredictable power production from RES, flexible storage, and controlled or uncontrolled power consumption. This is valid for the existing status and for the future viewpoints encompassing large-scale of RES with the intrinsic irregular and unpredictable power production [30], as well as taking into account the user demand profile for optimizing the power consumption exclusively from RES [31]. Although the investigated control methodologies to deal with the power production from RES, the presence of ESS offers new possibilities of power management, also requiring dedicated control methodologies [32].

The vehicle electrification is also recognized as vital for a cooperation control between RES and ESS [33, 34], supporting the reduction of energy costs and greenhouse gas emissions and commit for a cooperative power optimization [35–40]. This cooperative scenario is pertinent when framed with smart grids and also with smart homes [41–44], where the scheduling uncertainties of the EV is also an issue that must be considered, targeting to enhance the grid performance [45–48]. Moreover, advanced topologies for simplifying and unifying RES and EVs are also fundamental [49].

Despite the advantages of the vehicle electrification for the smart grids, its impact on the electrical grid operation is of utmost importance and must be handled properly [50–53]. An on-line adaptive strategy for coordinating the EV parking schedules, in the perspective of maintaining the operability of the electrical grid, as well as the user convenience, is proposed in [54]. Similarly, an approach to minimize the peak loads of the electrical grid and the EV charging costs at the same time is proposed in [55] for a coordinated integration of the vehicle electrification. Concerning the analysis about the impact of the vehicle electrification in electrical grids, exemplification research works developed around the world are presented, respectively, in [56–60], for the Australia, Canada, China, Sweden, and Portuguese cases.

As the title specifies, this chapter deals with challenges and a global perspective of the vehicle electrification in smart grids. Contextualizing the above-mentioned subjects, this chapter incorporates contributions and overviews in the following fields: Section 2 introduces the different structures concerning the internal constitution of an EV battery charger (EVBC) in terms of power stages, as well as its principle of operation; Section 3 summarizes the main communication technologies for the vehicle electrification, establishing different perspectives in smart grids and smart homes; Section 4 presents a global overview about challenges and opportunities of wireless power transfer in the perspective of the vehicle electrification; Section 5 discusses the relevance and the future perspectives about a direct or an indirect bidirectional power transfer between EVs, operation denominated as vehicle-to-vehicle; Section 6 introduces unified technologies for the vehicle electrification, permitting to combine the functionalities of an EVBC and a motor driver in a single equipment; Section 7 contextualizes the operation modes for the vehicle electrification and presents a set of opportunities offered for future scenarios of smart grids and smart homes. The book chapter is finalized with Section 8, where are presented the main conclusions according to each section.

2. EV battery chargers: an analysis of the principle of operation and of the power stages

An EV battery charger (EVBC) is classified either as on-board or as off-board, depending if it is installed inside or outside the EV, respectively. Regardless the on-board or off-board concept, internally, an EVBC incorporates power electronics converters with the respective control system. **Figure 1** illustrates an EVBC in its conventional structure, organized by two power stages: an ac-dc front-end interfacing the electrical grid and controlled by a current feedback; and a dc-dc back-end interfacing the EV battery and controlled by a voltage or a current feedback. The presented variables are the main required for a closed-loop control. This figure also shows the signals for the power stages. Although there are two distinct power

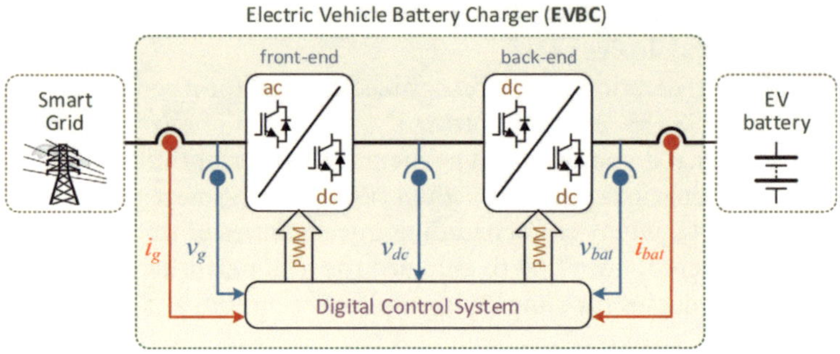

Figure 1.
EVBC constituted by two power stages (a front-end and a back-end, respectively, to interface the electrical grid and the EV battery) and by a common control system.

stages, each one with a specific strategy to ensure that current and voltage are precisely controlled, the control system should be viewed as a whole, since both power stages are linked by a dc-link. This is crucial to emphasize since, for example, the grid-side current of the front-end power stage is controlled according to the voltage and current levels of the battery-side of the back-end power stage. This means that the amplitude of the grid-side current is a function of the charging power in the battery-side. Therefore, a global power theory for the EVBC is applied for determining the reference of current for the front-end power stage [61, 62]. On the other hand, the reference of voltage or current for the back-end power stage is determined by the battery management system (BMS) [63, 64]. Based on the established references, individual and dedicated control strategies are applied for each power stage, basically, to determine the status of the switching devices during each control period [65–67] Besides the aforementioned low-level control requirements for both power stages (in terms of the switching devices), a communication platform, within the whole control system, is essential for establishing a bidirectional communication with the smart grid or smart home. The different technologies for the communication, as well as the functionalities framed with the EVBC, are analyzed in Section 2.

The conventional structure of an EVBC is based on two power stages, regardless the on-board or off-board concepts and the topology [68–70] Nevertheless, other structures are possible, for example, by combining these two concepts for an EVBC (this means that an EVBC can be constituted by an off-board power stage and by an on-board power stage) or a structure based on a single power stage. Analyzing the power stages in more detail, different arrangements are possible, for instance, based on multilevel structures [71–74] interleaved topologies [75, 76], and with or without galvanic isolation [77, 78] Independently of the arrangement, high-levels of power quality concerning low harmonic distortion, high power factor, and balanced currents (in the case of three-phase EVBC) must be guaranteed [79]. Similarly, a voltage and a current with low-ripple must be guaranteed for the battery-side in the perspective to preserve the battery lifetime. **Figure 2** shows the possible structures that can be implemented for an EVBC, highlighting on-board and off-board concepts, as well as power stages encompassing galvanic isolation.

Figure 3 illustrates an EV with the two possibilities of interfacing the electrical grid: an on-board and an off-board. As demonstrated, the power stages permit a bidirectional power flow, from the electrical grid to the EV and vice-versa. This possibility is denominated as grid-to-vehicle (G2V) or, in reverse, denominated as vehicle-to-grid (V2G). Dedicated control algorithms are responsible for controlling the EVBC in one of these modes, where the power management is accomplished by the smart grid or by the smart home. Moreover, the EV user has also some privileges (for instance, financial incentives) when allowing the EV operation in a flexible controllability of G2V/V2G modes, but without ignoring the information provided by the BMS.

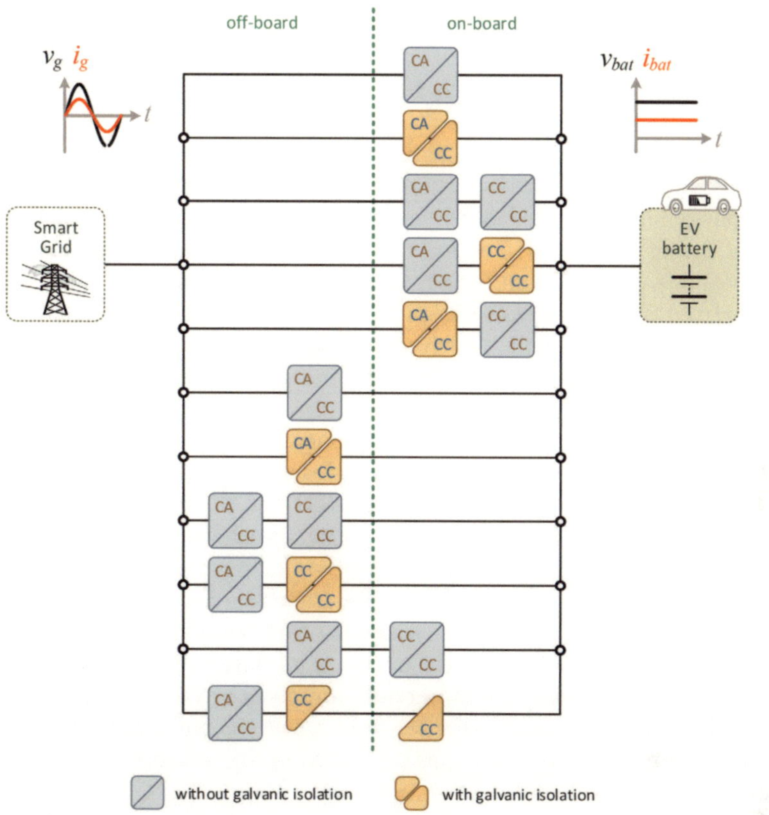

Figure 2.
Possible structures that can be implemented in EVBC, highlighting on-board and off-board concepts, as well as power stages encompassing a galvanic isolation.

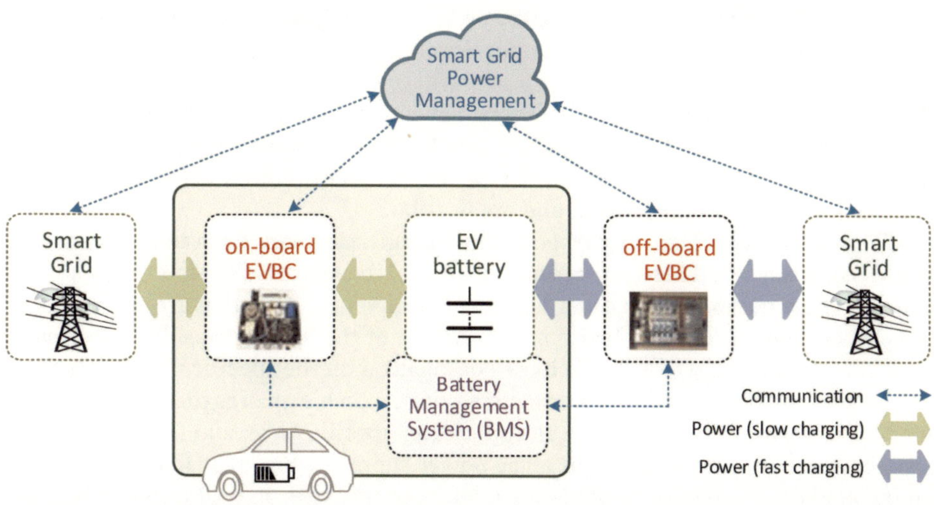

Figure 3.
EV interface (through an on-board and an off-board EVBC) with the smart grid, and establishing bidirectional communication and bidirectional power flow.

The same way as for an on-board EVBC, and also permitting a bidirectional exchange of active power from the electrical grid to the EV and vice-versa, is the operation of an off-board EVBC. Moreover, in terms of controllability, the same principle is applied by combining the requirements and benefits of the EV user, the battery BMS, the smart grid, and the smart home.

3. Communication technologies for vehicle electrification

This section presents an overview of communication network standards in the context of vehicle electrification. A communication network is a communication system that enables bidirectional communication between the multiple devices that are part of the network (called nodes). This is in contrast with broadcast communication systems (e.g., TV and radio), which only allow communication on one direction (from the station to the users' devices), and with bidirectional point-to-point communication systems, which only allow communication between two devices. In order to be able to communicate, the network nodes must use the same protocols. In this sense, the use of standard network technologies and protocols is normally preferable due to criteria such as compatibility, variety of choice and cost. The term "network" may apply to a set of nodes that exchange data using a specific network technology and its respective protocols (e.g., Wi-Fi or Ethernet) or may be used in a broader sense to refer to a communication system that interconnects devices that communicate using different networks technologies and protocols (e.g., the Internet). Most of the traffic on the Internet (e.g., audio, video, and data files) is produced and consumed by humans. In contrast, the concept of IoT [80, 81] extends the use of the Internet for the exchange of data generated, processed, stored, and analyzed by devices, either autonomously or with the participation of the users. Given the relevance of the IoT for vehicle electrification and smart grids [82], the communication technologies presented in this section are framed in the scope of the IoT.

3.1 IoT architecture

The IoT paradigm is also being deployed in several other contexts and applications areas, including different industry sectors [83, 84], smart cities [85, 86], and healthcare [87, 88]. Besides data communication, an IoT system has to perform a multitude of other tasks. Therefore, it is useful to organize these tasks in an IoT architecture composed of different layers. Several authors proposed different architectures [80, 84, 89, 90], but there is not yet a consensus regarding a reference model. A basic proposal that is suitable for this chapter divides the IoT architecture in three layers: sensing, network, and application.

The sensing layer is mainly responsible for collecting data from the physical world using sensors. These sensors are integrated in electronic devices (sensor nodes). These nodes include other hardware components [91] that are essential for the proper operation of the device in the context of the IoT, such as: (i) a communication transceiver, which needs to be compliant with the specific network technology used by the device; (ii) a processing unit, which executes the software for the higher network layers, as well as application-specific code; and (iii) a power source, which may be a battery or an ac power supply (when available), depending on the application requirements. Some devices of this layer may also integrate actuators, which perform an opposite role compared to sensors, acting on the physical world based on the received data. The sensing layer may also be called as perception layer [92, 93] or objects layer [80].

The network layer provides the interconnection between the sensing layer and the application layer. This layer normally is composed of multiple types of communication networks, which form the communication infrastructure used for the exchange of information between the different types of devices that are part of the IoT. An overview of relevant network standards is provided in the next subsection. This layer also handles other IoT tasks, such as data storage and cloud computing [92]. In this sense, several IoT cloud platforms [94] are currently available, as well as cloud computing platforms from major players that also provide IoT services, such as Amazon Web Services (AWS), Google Cloud Platform, Microsoft Azure, and IBM Bluemix.

The application layer is the higher level of the IoT architecture. This layer provides specific services for the users based on the data collected by the sensing layer. These services include the automation of processes, using application-specific control algorithms designed in the scope of this layer. In the context of vehicle electrification applications, one example is the EVBC control, inside a smart home, based on the measurement of the total instantaneous home current, in order to avoid the tripping of the main circuit breaker [95]. The services that may be performed by this layer include also the provision of a user interface through an IoT client device, to allow the user to interact with the IoT system [96]. This layer also includes data-mining algorithms [97, 98].

3.2 Network standards

The devices that are part of an IoT system may range from multiple distributed sensor nodes, at the lower level, to centralized cloud computing servers, at the higher level. These devices present different requirements and capabilities in terms of data rate, energy consumption, processing power, connectivity, etc. For example, many applications require the deployment of several low-cost wireless sensor nodes to collect relevant data [99]. Without cables, the sensor node has to resort to a battery as its power source; therefore, it is normally designed to operate with very low energy consumption, in order to maximize its lifetime. Typically, these nodes also require low data rate and offer low processing power. On the other hand, a cloud server normally requires a high data rate wired connection and machines with high processing power and high energy consumption to handle the data collected from multiple devices. Therefore, the network layer of the IoT architecture requires different network technologies, organized hierarchically from lower to higher levels, in order to satisfy the requirements of its different IoT devices.

Wireless networks may operate in unlicensed or licensed frequency bands. The unlicensed bands were reserved originally for radiofrequency (RF) emissions of industrial, scientific, and medical (ISM) equipment for purposes not associated with communications (e.g., microwave ovens). Nowadays, ISM bands are used by short range wireless networks such as ZigBee, Bluetooth, and Wi-Fi, as well as some low power wide area networks (LPWAN), such as LoRa. The main advantage of ISM bands is that they can be used without a government license. On the other hand, they may be subjected to interference from other wireless communication devices and ISM equipment. Normally, there are multiple channels in these bands, so the network devices may select channels with less interference for operation. The main ISM bands currently used by wireless networks are the 433 MHz, 900 MHz, 2.4 GHz, and 5 GHz bands, but the first two are not available worldwide. Higher frequency bands tend to have more bandwidth available, which means that the wireless networks may offer higher data rates. On the other hand, lower frequency bands allow longer range. The channels in licensed frequency bands are normally sold by the government to operators, which offer their services to their users (e.g., mobile cellular network operators). These channels suffer less interference due to exclusive allocation to a single operator. However, the use of these wireless networks normally has costs to the user.

Wireless networks may also be classified according to their range. In this sense, short-range networks include personal area networks (PAN) and local area networks (LAN), whereas long-range networks include wide area networks (WAN). Wireless PAN (WPAN) standards [100] that are suitable for IoT include ZigBee [101], Bluetooth Low Energy [102].

The two lower layers of the ZigBee protocol stack, physical (PHY) and medium access control (MAC), are defined by the IEEE 802.15.4 low power and low data rate WPAN standard [103]. The PHY layer uses direct sequence spread spectrum (DSSS) and offers PHY data rates up to 250 kbps. There are 16 channels in the 2.4 GHz ISM band, available worldwide, as well as 11 channels in the 868/915 MHz bands, but these are available only in some regions of the world. The MAC layer is based on a CSMA/CA (Carrier Sense Multiple Access/Collision Avoidance) algorithm. ZigBee defines three types of nodes: coordinator, router, and end-device. The ZigBee coordinator starts the network formation and selects the communication channel, among other tasks. It may also perform the same tasks of ZigBee routers, which include routing of packets between nodes and allowing other nodes to join the network. The ZigBee end devices are located at the extremities of the network, which means that they cannot route packets. On the other hand, the end devices may sleep to save energy, making them suitable for battery-operated sensor devices. Although the direct range of ZigBee devices is limited to dozens of meters, the support of multi-hop mesh topology (**Figure 4a**) allows increasing the network range. ZigBee can also operate in star topology (**Figure 4b**), which is the topology normally used also by most of the other wireless networks, such as Bluetooth, Wi-Fi, and mobile cellular networks. In [104], the authors discuss the use of ZigBee and other wireless technologies in the context of intelligent transportation systems (ITS).

Bluetooth low energy (BLE) was introduced in the Bluetooth 4.0 specification. It operates in the 2.4 GHz ISM band using frequency hopping spread spectrum (FHSS). The original standard provides a PHY data rate of 1 Mbps. The MAC protocol is based on a master-slave polling mechanism. BLE was developed for use in low power wireless devices, such as battery-operated sensor nodes. BLE is not

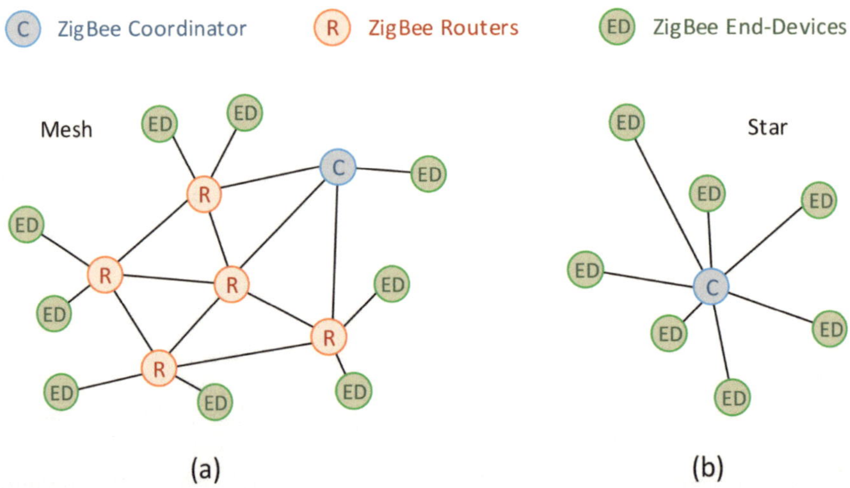

Figure 4.
Main ZigBee topologies: (a) mesh (multi-hop); (b) star (single-hop).

compatible with classic bluetooth, which continues to be offered for other applications (e.g., audio transmission). The new Bluetooth 5.0 specification introduces improvements in terms of data rate and range, among others [105]. BLE was designed to operate in star topology, but there are proposals to increase its range though the use of BLE mesh networks [106]. In [107], the authors present the development and test of a BLE network for wireless monitoring and control of parameters associated to the battery and traction systems of an EV.

WLAN technologies normally provide longer range and much higher data rate than WPANs, but they also tend to have higher power consumption. Although there were other WLAN alternatives in the past, such as high performance radio LAN type 2 (HIPERLAN/2), the WLAN market nowadays is dominated by Wi-Fi products. Similarly to ZigBee, the PHY and MAC layers of Wi-Fi networks are defined by IEEE standards, in this case, of the IEEE 802.11 family [108]. The original IEEE 802.11 standard defined PHY data rates of 1 and 2 Mbps in the 2.4 GHz ISM band. The IEEE 802.11b amendment increased the maximum data rate to 11 Mbps. IEEE 802.11a/g/n/ac/ax amendments extended the operation to the 5 GHz band and increased significantly the data rate, through the use of wider channels and higher-order modulation techniques. Several other amendments were specified, with improvements in other areas. For example, IEEE 802.11p defines enhancements to support vehicular networks [109], in the scope of ITS, including vehicle-to-vehicle (V2V) and vehicle-to-infrastructure (V2I) communication and operating in specially allocated licensed ITS bands at 5.9 GHz.

Concerning wired network technologies, some alternatives available to provide local area communication in the context of IoT systems are IEEE 802.3/Ethernet [110, 111] standard and power line communication (PLC) technologies [112, 113].

When longer ranges than the ones provided by WPANs and WLANs are needed and the data rate requirements are low, the use of a low power wide area networks (LP-WAN) [114], such as LoRa [115], Sigfox [116], or Narrowband IoT (NB-IoT) [117] may be considered a better alternative. In [118], the authors propose an EV charging architecture based on the use of LoRa networks.

WPANs and WLANs are normally only suitable to provide local communication from sensor nodes to a nearby base station (e.g., Wi-Fi access point, ZigBee coordinator, or BLE master). One example is the use of these wireless network technologies for indoor communication inside a smart home. Therefore, it is also necessary to use other communication technologies to transfer the collected data from the base station to the IoT servers through the Internet, using appropriate wired and/or wireless wide area networks (WAN). Normally, this connection is provided by an Internet service provider (ISP) operator, using digital subscriber line (DSL) over twisted pair, coaxial cable, or fiber optic. An alternative is the use of cellular network technologies.

3.3 Higher layer protocols and gateways

Communication networks are normally structured into five protocol layers: physical, data link (or MAC, in wireless networks), network, transport, and application [119]. In order to communicate directly with other devices on the Internet, a sensor device needs to implement the higher layer protocols of the Internet protocol suite. This means that the device needs to implement Internet Protocol (IP), at the network layer (either IPv4 or IPv6). At the transport layer, there are two main options: transmission control protocol (TCP) or user datagram protocol (UDP). TCP provides error correction through retransmissions, whereas UDP is a lightweight transport protocol that provides only error detection, which means that error correction mechanisms have to be provided by other layers if required.

Application layer protocols include hypertext transfer protocol (HTTP), message queuing telemetry transport (MQTT), and constrained application protocol (CoAP) [120]. While HTTP is widely used on conventional Internet applications, MQTT and CoAP are lightweight protocols more suitable for IoT applications.

These higher layer (network, transport, and application) protocols are normally implemented in Wi-Fi devices, together with the specific PHY and MAC lower layer layers specified by the IEEE 802.11 standards [121], allowing seamless communication in IoT applications, as well as the sharing of the Wi-Fi network with the conventional Internet traffic. In contrast, IEEE 802.15.4/ZigBee and BLE implement their own higher layer protocols, which are optimized for low power devices and not directly compatible with the Internet protocol suite. The 6LoWPAN protocol, which compresses the IPv6 header to make it more suitable for low power wireless networks [122], was designed to enable direct connection of IEEE 802.15.4 networks to the Internet.

An alternative to allow the integration of these WPAN devices into the IoT is the use of a gateway device to translate the packets exchanged between the WPAN and the Internet. The same gateway may also be used to provide other functionalities, such as: a local database; a controller node, running automation algorithms associated to the IoT application; a security firewall, monitoring and controlling the communications in order to protect the WPAN devices from malicious attacks [96]; or a MQTT broker. An example of a BLE/Wi-Fi gateway implementation based on a Raspberry Pi 3, which also acts as local database, is provided in [123]. In some applications, it may also be interesting to use a smartphone as a gateway, especially in mobility scenarios. An example is described in [124], where a smartphone is used to provide connection between the nodes of a BLE-based intra-vehicular wireless sensor network (IVWSN) and a Google Firebase database using 4G and Wi-Fi.

3.4 Related work

As discussed before, there are many types of communication technologies that can be selected, based on their characteristics (such as communication range, transmission data rate, energy consumption, data transmission costs, mobility, etc.), to provide a suitable communication infrastructure for a particular application scenario. This section provides guidelines and examples of how these communication technologies can be used in the context of vehicle electrification applications.

The G2V/V2G collaboration for the smart grid reliability, analyzed in the perspective of communication is presented in [125], where an extension to the IEC 61850–7-420 standard is proposed as a support for coordinating the EV in G2V/V2G modes in terms of an information model.

In the perspective of the EV integration into smart grids, bidirectional communication between the EV and roadside units (V2I) is also fundamental. In this context, simulations and a comparative analysis in terms of communication performance between the EV and roadside units are conducted in [126], where wireless communications technologies as ZigBee, Wi-Fi, and Worldwide Interoperability for Microwave Access (WiMAX) [127], were considered. In the comparison, the results were mainly focused on the doppler effect and the end-to-end delay, taking into account the requirements of the IEC 61850 and the IEEE 1609 WAVE standards.

The integration of V2I and V2G communication with smart grid components such as CSs using different information models based on the IEEE 1609WAVE and IEC 61850 standards is investigated in [128]. The assessment is based on the evaluation of the end-to-end delay over diverse vehicular ad-hoc network protocols.

Simulation results show that the protocols with lower overhead are able to achieve better performance.

As demonstrated in [129], the IEC 61850 communication standard can also be used for the energy management (EM) of EVs in microgrids, based on smart algorithms for the G2V/V2G modes and distinct modes (as example, valley mode, steep hump mode, flat and low hump mode). The proposed method is based on the extensible messaging presence protocol (XMPP), and its mapping to the service models is demonstrated as a solution for the energy management problem, providing network security and scalability.

A IEC61850-based communication system, in the perspective of the power management within a smart home with an EV in G2V/V2G mode and with RES (PV in this case), is proposed in [130], where the management structure is supported by the estimation of the RES production and the EV battery state-of-charge. The communication messages were transmitted by means of different wired and wireless communication technologies, and the presented results demonstrate that the delays are within the limitations imposed by IEC 61850 standard.

4. Vehicle electrification: a comprehensive perspective of wireless charging systems

EVs need electricity to run their motors. This electricity either can be supplied by an on-board battery, which must be periodically recharged from the electrical grid, or can be directly obtained from a continuous over time connection to the power grid itself. Variations around this classification can be made, as electric energy can be generated, for instance, by a thermic engine installed on-board of a hybrid vehicle, or it can be delivered to the vehicle by an off-grid power plant. The point is that electricity either can be originated from an energy source or storage that is co-installed on board of the vehicle, or can be delivered by a stationary power plant, which is external to the vehicle, as shown in **Table 1**. In the first case lie almost all passenger cars so far seen in the market, whereas in the second case are trains and trams, which are energized either by rails or overhead wires and, commonly, ride on tracks. The tracks mechanically restrict the lateral displacement of the vehicle, what helps keeping the alignment of the vehicle to the electrified rail or the overhead line, an essential condition for the power transfer to occur. A vehicle with a self-contained energy storage will still normally need external electrification, at least during the stationary charging cycle.

Until very recently, the only way for an EV to get its battery recharged was by wired transference of electric energy, that is, by galvanic contact. If for some safety reason galvanic isolation, between the EV and the electrical grid is required, an isolating transformer should be employed in the charging station. After wireless power transfer (WPT) advances have entered the market of cell-phone recharging and small, low-power, home electrical appliances, the automotive industry is now the major target: The recent development of WPT technologies now make it possible to obtain stationary EV charging stations without cables or any galvanic

Vehicle kinematics	Self-contained energy storage	External electrification
Vehicle immobilized	Idle/standby charging	Stationary charging
Vehicle in movement	Discharging/in-march charging	Dynamic electrification/charging

Table 1.
Types of vehicle electrification.

electric contact between the EV and the charging unit, which can be hidden underneath the floor surface. This is now gaining commercial status and, in the near future, many units are expected to be seen (or more precisely, not seen, for they can be concealed in the floor) in garages and parking lots.

A novel WPT-based system for the vehicle electrification with an active power of 11 kW is presented in [131], where special reflections were considered for the misalignment between the WPT: primary coil (off-board the EV) and secondary coil (on-board the EV). A 10-kW WPT prototype dedicated for EVs is proposed and validated in [132], obtaining an efficiency of 94%. In a global perspective, an analysis of the state-of-the-art of WPT technologies, as well as a review of industrial projects under development, is presented in [133]. An overview about WPT technologies, as an influence for a sustainable mobility, is offered in [134], including sustainable performance, technical progresses, and applications of WPT. Complete overviews concerning WPT technologies focusing in electric mobility applications are presented in [135–137].

There are also multiple ongoing research efforts to make WPT a viable technology choice for dynamic electrification. Currently, built prototypes of electrified pathways are capable of sustaining 20 kW of electric power dynamically delivered to a moving vehicle on a 100 m long road segment [138]. However, even with the great advances in materials and power electronics, the technology is still expensive and not fully engineered to large-scale applications.

In this section, the principles of wireless power transfer and some of its automotive applications mostly focusing charging systems are reported, pointing out new achievements in the field.

4.1 Wireless power transfer

The application of WPT technology is reported to have been envisioned far behind, in late nineteenth century, by Nikola Tesla. WPT is based on two phenomena relating magnetism and electricity. The first was discovered by Hans Christian Ørsted in 1820 and theorized by André-Marie Ampère shortly after [139], and consists in the creation of magnetic field by electric currents. The second, the induction of electric by varying magnetic fields and the existence of mutual induction between two windings magnetically coupled, was later demonstrated by experimentation and theorized by Faraday [140], in a sequence of experiments starting in 1831. These results were later integrated in a treatise by Maxwell [141], which was later simplified by Oliver Heaviside to what is currently known as the Maxwell's Equations [142].

The applications of Maxwell's Equations to magnetically coupled coils lead to simplified models of transformers that were extensively proved by experimentation and that can effectively allow the calculation of energy transfer over these entities with circuit theory. In this way, the behavior of magnetically coupled coils, as shown in **Figure 5**, under harmonic excitation at a low enough frequency (so that the system can be considered not to irradiate energy) and negligible resistive losses, is well described by the simplified transformer model in **Figure 5** and the complex Eqs. (1) and (2).

$$V_1 = jwL_1.I_1 + jwM.I_2 \tag{1}$$

$$V_2 = jwL_2.I_2 + jwM.I_1 \tag{2}$$

The power P transferred over the two magnetically coupled coils is then given by (3):

DOI: http://dx.doi.org/10.5772/intechopen.89655

$$P = V_{M_2}. - I_2^* = -jwM.I_1.I_2^*, \quad w = 2\pi f \tag{3}$$

From (3), it can be inferred that the power wirelessly transferred through the coils is proportional to the working frequency f, to the mutual inductance M, and the RMS currents $|I_1|$ and $|I_2|$. But, the power also depends on the relative phase between currents I_1 and I_2: If they are either in phase or in counter-phase (180° apart), no real power is exchanged between primary and secondary, only reactive power being involved. For I_1 and I_2 of fixed module, the transferred power from primary coil to secondary coil is maximized when these currents are 90° out of phase (4):

$$I_2 = -j.I_1 \tag{4}$$

In order to adjust the intensity of primary and secondary currents I_1 and I_2 and keep them as close as 90° as possible, so that condition (4) is observed, impedance compensation circuits should be added to the primary and secondary coils, as in **Figure 6a**. One of the possibly simplest compensation circuits, and the first to be used in Tesla's experiments, is the pure series capacitive compensation. The configuration derived when series capacitive compensation is employed in both primary and secondary, is called the series-series (SS) impedance compensation, shown in **Figure 6b**.

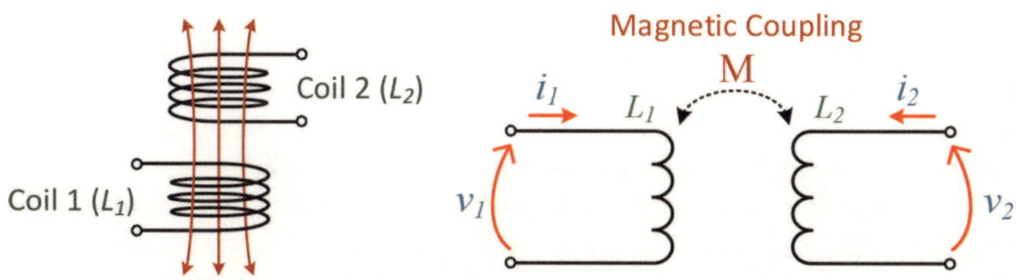

Figure 5.
Magnetically coupled coils and the equivalent transformer model in circuit theory.

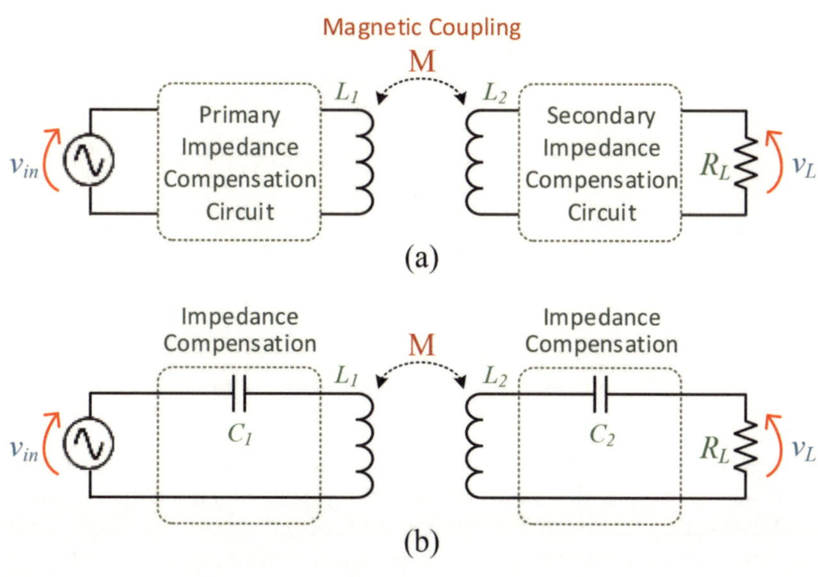

Figure 6.
Basic WPT configuration: (a) exemplified with the series-series (SS) impedance; (b) impedance compensation.

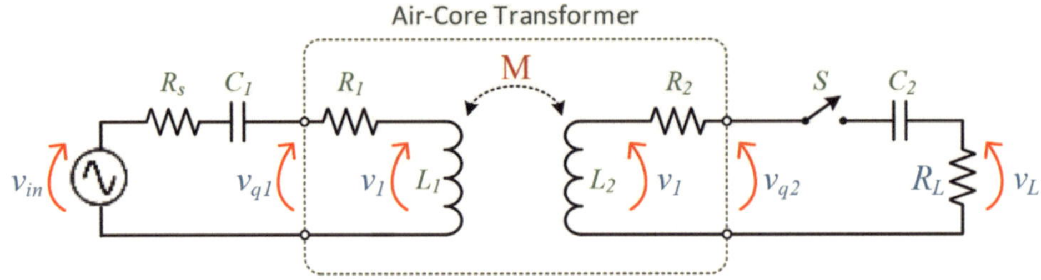

Figure 7.
Simplified circuit analysis for the SS-compensated WPT.

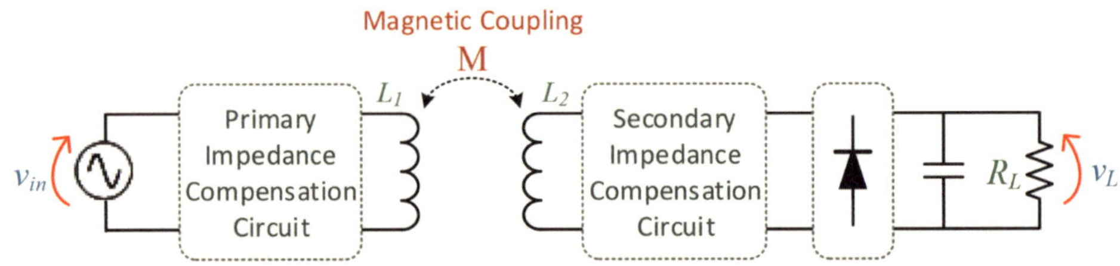

Figure 8.
Basic fixed gap WPT schema for dc load, with voltage rectification in the secondary.

The model in **Figure 5** is too simplified because no power losses in the winding or elsewhere are considered. When using the SS compensation in WPT the improved model that is still simple and is still able to represent the losses in windings of the primary and secondary coils is shown in **Figure 7**.

In this circuit, R_1 and R_2 are respectively the total series resistance of the capacitor and the inductor, in primary and secondary circuits, R_s is the impedance of the power source exciting the primary circuit and R_L is the load consuming the net power transferred to the secondary circuit. By circuit analysis, it can be derived that the electrical efficiency η of the power transfer scheme, from primary to secondary coils, at the resonance condition (5).

$$2\pi f_0 = \frac{1}{\sqrt{L_1 C_1}} = \frac{1}{\sqrt{L_2 C_2}} \tag{5}$$

is given by (6) and (7):

$$\eta = \frac{1}{\left(1+\frac{R}{R_L}\right)\left(1+\frac{1}{k^2}\left(\frac{R_s+R_1}{w\,L_1}\right)\left(\frac{(R_2+R_L)}{w\,L_2}\right)\right)} = \frac{1}{\left(1+\frac{R}{R_L}\right)\left(1+\frac{1}{k^2 Q_1 Q_2}\right)} \tag{6}$$

where:

$$Q_i = \frac{w\,L_i}{r_i}, \qquad i \in \{1,2\}, \qquad r_1 = R_s + R_1, \qquad r_2 = R_2 + R_L \tag{7}$$

and f_0 is the frequency of the power source V_{in} exciting the primary circuit. An equivalent algebraic formulation for the efficiency η is given in [135]. The factors Q_i are called the quality factors of the primary and secondary coil windings. Commonly, the load connected to the secondary requires dc voltage, so a voltage rectification and stabilization circuit is required in the secondary, as exemplified in **Figure 8**.

Advances in the study of WPT models indicate the better adequacy of new impedance compensation topologies in automotive applications, such as the inductor-capacitor-capacitor (LCC) circuit, for both primary and secondary coils, as reported in more recent work [143, 144].

4.2 Stationary WPT charging

Based on so far available knowledge on WPT charging for light duty vehicles, the Society of Automobile Engineers (SAE) issued in 2016 a general recommendation for stationary WPT in automotive applications [145]. A simplified cross-section of a typical coil-to-coil WPT assembly is shown in **Figure 9**.

It can be seen in **Figure 9** that the coils are placed parallel and center-aligned to each other, with ferrite plates around them to increase the mutual inductance, as the amount of transferred power, according to (3), is proportional to this parameter. Parallel aluminum plates partially enclose the coils, as to function as a magnetic shield, reducing the magnetic field that spreads outside the gap in between the coils. A set of recommendations concerning the geometry of this assembly was also included in the same document [145], the SAE J2954 Report, and is concerned with the future interoperability of stationary recharging equipment for the automotive industry. This document, which is due to eventually evolve to an industry standard, also predefines three power levels classes and a frequency operation band for the WPT, as shown in **Table 2**.

The current available technology strongly limits the maximum distance in between the primary and secondary coils that can be achieved at a reasonable electric efficiency, for the given desired power levels. The SAE J2954 also establishes

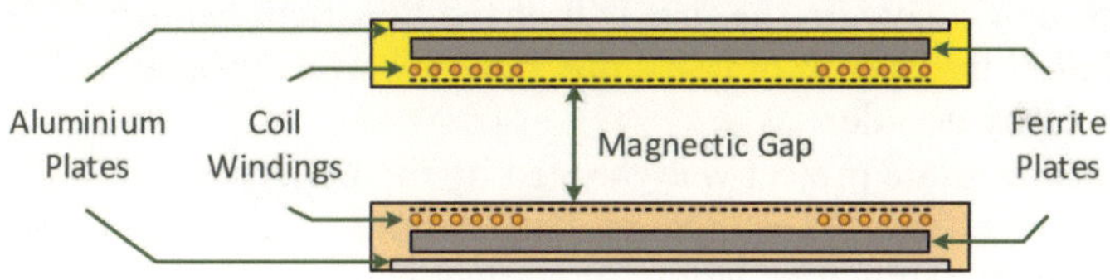

Figure 9.
Typical cross-section of the coil-to-coil WPT assembly of stationary chargers.

SAE J2954 recommended practice (as of Nov 2017)	WPT power classes		
	WPT1	WPT2	WPT3
Maximum input power	3.7 kW	7.7 kW	11 kW
Frequency band	81.38–90 kHz		
Transfer efficiency	>85% @ full alignment		

Table 2.
WPT power levels for stationary automotive charging according to SAE J2954 recommended practice.

SAE J2954 Z-class	Ground clearance range (mm)
Z1	100–150
Z2	140–210
Z3	170–250

Table 3.
SAE J2954 ground clearance range as per defined Z-classes.

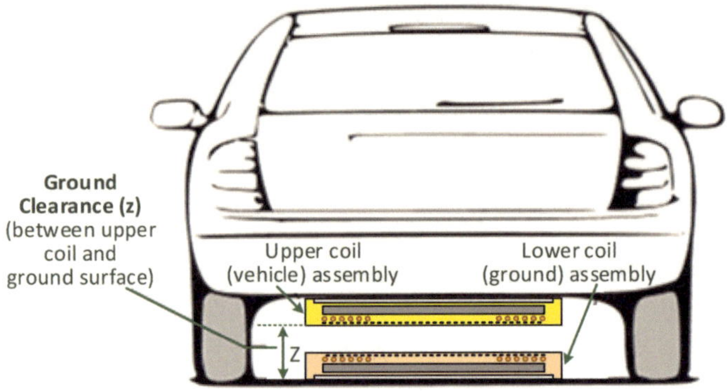

Figure 10.
Typical configuration for stationary WPT, showing coil in vehicle assembly (yellow) in alignment with coil in ground assembly (orange).

classes of possible clearances between the coil installed in the vehicle and the ground, what ultimately defines the gap between primary and secondary coils. SAE J2954 ground clearance range as per defined Z-Classes (**Table 3**). The ground clearance ranges by these named SAE classes, named Z1–Z3, are given in **Figure 10**.

4.3 Dynamic WPT electrification

The first automotive WPT designs targeted a means of dynamic electrification of vehicles, not stationary charging. The motivation was to minimize battery capacity requirements, not stationary charging: The subject was brought to light by George Babat, in Moscow, in the first half of the twentieth century [146, 147]. Due to many technical limitations of that time, however, the dynamic WPT remained forgotten for many decades.

In 1979, a conceptual project was charged by the University of California at Berkley to Systems Control Technology, Inc., Palo Alto, CA, USA [148]. The conception of a dynamic inductive WPT system was completed by 1986, when E.H. Lechner, S.E. Shladover, and K. Lashkari published two articles in the 8th International Electric Vehicle Symposium, Washington D.C. [149, 150], reporting the design of a Roadway Powered Electric Vehicle (RPEV). The final demonstration version of the system consisted of a 213 m long inductive road segment that could power an electric bus demanding 48 kw at an average electric efficiency in the order of 50–55%, and up 60% in peak conditions. In spite of the good qualities of the concept, practical limitations of power electronics components of the time influenced the design to be implemented at the low frequency of 400 Hz, with intensive use of iron alloy cores for the magnetic links in between the road and the vehicle, thus resulting in an unattractive cost level, so that the idea was abandoned [148].

In the 1990s, the interest on WPT for automotive applications was definitely recovered with the work of Covic and Boys [151] and, since then, many efforts in this direction have been pursued, with the most representative of them being the FABRIC project [138], as referred in [4]. In this design, a 100-m long track was built to deliver 20 kW of power, to up to two vehicles simultaneously running over it. The complexity and cost of enterprises like this are still too high for widespread adoption, but this is a topic of current research interest and it is believed that some kind of dynamic WPT will eventually become popular. In a general form of dynamic WPT implementation, the distance and relative orientations between primary and secondary coil are assumed to vary in a certain range. This relative

DOI: http://dx.doi.org/10.5772/intechopen.89655

movement will cause dynamic variations in the magnetic coupling between the coils and, potentially, variable self-inductances of both primary and secondary coils as well, what will continuously change the transfer function of in between the coils, affecting the resonance frequencies exhibited by the whole assembly.

A possible solution to keep the power flow about constant is to allow both the excitation frequency and power level delivered to the primary coil to be also dynamically adjusted. In terms of circuit equivalence, dynamic WPT can then be modeled as in **Figure 11**. A generic wireless data channel is also illustrated, which is used to receive feedback from the secondary-side and to enable the control of the primary-side (i.e., the v_{in} voltage in terms of amplitude and frequency).

However, the dynamic WPT has also some disadvantages that must be viewed as challenges for future applications of this technology. A key challenge is the misalignment that can occur between the primary and secondary, which inevitably tends to deteriorate the transferred power. A homogeneous WPT technology targeting an effective dynamic WPT with moving objects is proposed in [152], where an experimental verification is demonstrated. A dynamic WPT containing numerous primary coils (stationary in the ground-side) and an EV with a secondary coil (moving EV) is proposed in [153], where a downscaled 3 kW prototype is presented allowing to confirm the dynamic WPT with its principle of operation.

4.4 Electromagnetic field exposure control

The use of electricity always brings some risks that should be carefully controlled at system design phase and further diminished by the elaboration of operational norms and procedures. In wired (galvanic) charging, for instance, careful dimensioning of cables and connectors should be done, for there is always the risk of overheating or sparks, which can cause a fire or, depending on the environmental conditions, even an explosion. Modern wired charging systems, for instance, avoid sparks by only switching a power circuit electrically after steady mechanical contact guarantees a stable galvanic connection. The risk of electrocution is one more issue, especially under mechanical failure of connectors, and it is aggravated when the contacts or the floor are wet.

In WPT systems, most of these risks involved in wired chargers are not present, because the user does not have direct contact with electric power cables, plugs, or receptacles. However, WPT systems are wireless only in the sense that there are no cables connecting the charger unit and the vehicle. Internally, these units are also replete of cables and wires, which should be well dimensioned, isolated, and constrained from direct human contact much in the same way wired systems are. Also, the strong EMF generated by the WPT coils can induce eddy currents in

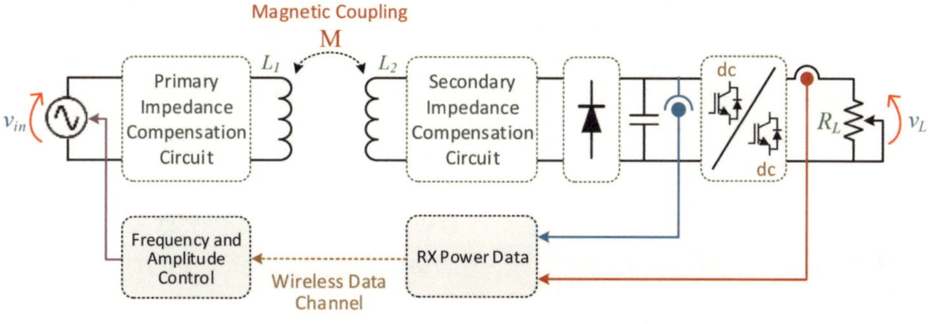

Figure 11.
Simplified dynamic WPT schema for dc load, with voltage rectification and stabilization in the secondary.

nearby resistive materials, which will heat and can start a fire. Similar happening with ferromagnetic materials, which can exhibit energy losses in the form of heat due to the alternating magnetic field. It is not enough to design for avoidance of such materials in the area exposed do the EMF. Monitoring the unexpected entrance in the WPT zone of objects made of such materials, the so-called "foreign objects," is also essential. A screw or nail stuck in a piece of wood left over a WPT transmitter can potentially start a fire. The system must then be able to automatically turn off an ongoing WPT and alarm, whenever foreign objects are detected. Additionally, the high intensity of EMF produced in the vicinity of both the transmitter and receiver coils can endanger human health. The level of human exposure to magnetic and electric fields tends to be much higher in WPT than in wired chargers. It is then necessary to carefully limit, by design, the maximum EMF in the area of human occupancy.

4.4.1 ICNIRP recommendations

As the knowledge regarding the long-term effects of EMF over human being progresses, the International Commission on Non-Ionizing Radiation Protection continually updates recommendations that are generally accepted by the society and the industry as de facto standards. This affects all engineered devices, including those in the automotive sector. The maximum International Commission on Non-Ionizing Radiation Protection (ICNIRP) recommended electric and magnetic RMS field exposure levels are established as a function of the frequency of excitation. For the operation frequency band recommended in SAE J2954, which is from 81.38 to 90 kHz, the maximum exposure levels for the general public are given in **Table 4**.

4.4.2 Special EMF recommendations for automotive WPT applications

Beyond the generally worldwide accepted ICNIRP recommendation for maximum human exposure to EMF fields, the SAE J2954 extends the recommended safety levels by considering the case that humans in the automotive application may have an implanted medical device (IMD), such as cardiac pacemakers, in which case the AAMI/ISO 14117-2012 standard should also be applied. This requires the use of more tight limits for the magnetic field strength in regions 2b and 3, as depicted in **Figure 12**. In these regions, SAE J2954 further requires the RMS values of the magnetic field to be limited to 15 μT and the peak magnetic field to 21.2 μT, in the adopted bandwidth for the automotive stationary WPT.

SAE J2954 still admits that conformity may still be observed if this additional requirement is not met, but in this case, steps should be taken to warn pacemaker wearers to avoid this region, that is, to stay away from the car. Since notices on the laterals and panel of the vehicle, warning that IMD users are under life threatening conditions are not very appealing for most drivers and passengers, in practice, this additional requirement must be observed. In region 2a (**Figure 12**), the basic

	General public exposure	Occupational exposure
Maximum electrical field strength	83 V/m	170 V/m
Maximum magnetic flux density	27 μT	100 μT

Table 4.
Maximum ICNIRP recommended EMF maximum RMS exposure levels to non-irradiating magnetic fields in the 3–10 MHz band.

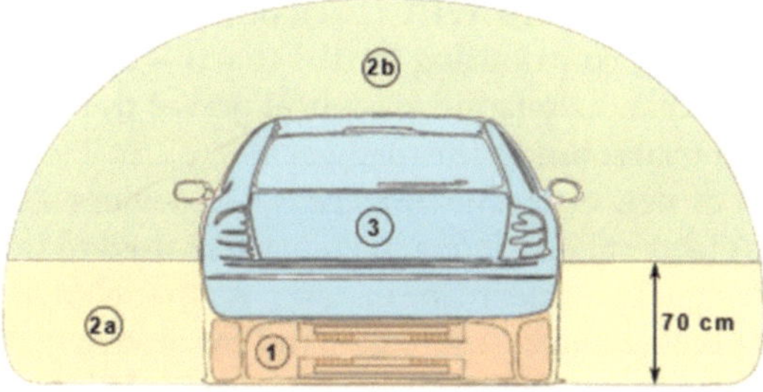

Figure 12.
EMF regions around a vehicle: 1. Space underneath the vehicle. 2a. Space outside the vehicle below the height of 70 cm from ground and excluding region 1. 2b. Space outside the vehicle above the height of 70 cm. 3. Space in the interior of the vehicle.

ICNIRP maximum recommended exposure level of upto 27 μT holds. In region 1, where the WPT phenomenon develops, no restrictions are posed – better not be there. As commented in [150], no conductive or magnetic debris should be left in that region as well, for the risk of overheating the parts and causing a fire. The recommendation and its referred norms are specific on the methods for measuring and verifying the field strength values.

4.5 New perspectives for WPT

In a future perspective, it is common sense that different types of EV will be available on market. In this sense, as previously demonstrated, with the possibility of the EV charging using WPT technologies, the main gains for the EV will be in terms of simplicity and comfort, since it is not necessary to use additional cables to maintain the EV plugged into the electrical grid. Moreover, it is an active approach to strength the market penetration of the vehicle electrification. A strategy to control the maximum power transfer points in WPT systems, based on arbitrary number of coils, is described in [154], where an experimental validation was performed at different modal frequencies and coils.

Many other innovative works have been accomplished in the last years: Compact dual-band WPT, constituted by two interlaced resonators, for instance, is proposed in [155]. It can operate in bidirectional mode, where a peak efficiency of 80% was obtained with an operating frequency of 300 MHz and considering a distance of 17 mm. In the perspective of WPT technologies framed in smart grids, a bidirectional WPT is of utmost importance, allowing to apply the G2V/V2G modes (for power transfer in both directions) with WPT [134]. An 1-kW bidirectional WPT prototype is proposed and validated in [156] focusing the resonant network in terms of active and reactive power control.

Since the efficiency is a key factor in WPT technologies, an innovative tracking method for guarantee maximum efficiency is proposed in [157], including aspects of: adjustment for coupling coefficient; variations of operating power; and controllability. Similarly, a system to guarantee optimum efficiency in WPT over a wide load range is proposed in [158].

Concerning the new technologies of WPT, the communication channels will also have a preponderant role. The main features concerning the communication protocols for WTP technologies, also based on wireless technologies, between the roadside controller and the on-board EV controller are discussed in [159]. The presented solution takes into consideration real-time aspects and the motion control. The combination of Internet of Things (IoT) communication networks with WPT technologies is explored in [160].

Besides the inductively-coupled WPT [161], other wireless power transfer technologies are also emerging, contributing for the fourth-generation of personal mobility [162]. As example, a dynamic via-wheel power transfer (V-WPT) is proposed in [163] as a trial solution for roadways.

The development of new electronic materials and devices allows continuous improvement in the electrical efficiency in converters used to implement WPT, simultaneously with significant decrease in costs. As the relative price of electricity with respect to fossil fuels reduces, more demand for electric vehicles and support technologies such as wireless power transfer is foreseen. While the advances in battery technology may eliminate the range anxiety of today's EV drivers, the batteries will always expectably need recharge. The WPT technology for stationary charging will be there for it.

Other grid and off-grid applications for the WPT stationary chargers and the electric vehicles themselves are expected to gain space among us, those related with vehicle-to-grid (V2G) and vehicle-to-home (V2H). In this sense, the new bidirectional WPT stationary charges will be able to dispose the electric vehicle as a mobile energy storage unit, extending by far its transportation functionality.

Last, it is worthwhile to mention that the lifecycle of batteries and the associated recycling issues, when projected at large scale, may perfectly justify the minimization of battery capacities by the widespread adoption of dynamic WPT on the roads as well.

5. Vehicle-to-vehicle: a power transfer perspective

As previously studied in this chapter, the EV interface with the electrical grid has a huge relevance for smart grids, considering the vast number of possible operation modes that the EV can allow. Besides the conventional modes G2V/V2G for exchanging active power with the electrical grid, there is a proposed operation mode in the literature termed as vehicle-to-vehicle (V2V). The V2V designation is mainly associated to communication systems between vehicles, either EVs or not. Nevertheless, a V2V operation mode considering power transfer between EVs was also proposed in the literature [164–167].

5.1 V2V power transfer using the front-end power stages

The V2V proposal intended to designate the power transfer between the batteries of different EVs connected to the same electrical grid, as a peer-to-peer power exchange method. Therefore, this operation mode is, in fact, the combination of G2V and V2G operation modes for two (or more) EVs connected to the same electrical grid, where the power provider operates in V2G and the power receiver operates in G2V. A practical implementation concerning a military environment was addressed in [168], where EVs would form a microgrid to replace diesel generators. Assuming that each on-board EVBC contains a front-end power stage and a back-end power stage, the power transfer from one EV battery to the other requires four power conversion stages. Thus, even if each power stage is highly efficient, the overall efficiency of the power transfer will always be lower than the least efficient converter. For instance, if all the converters have an efficiency of 90%, the overall efficiency of the power transfer would be 65.6%. Additionally, the power transfer between EV batteries is only possible if the EVs are connected to the same electrical grid. **Figure 13** illustrates this case of power transfer between EVs connected to the same electrical grid. Besides this possibility, where both EVs are plugged-in into the electrical grid and the front-end power stages are controlled by current feedback, other possibility consists in using one of the EVs as a voltage source and the other

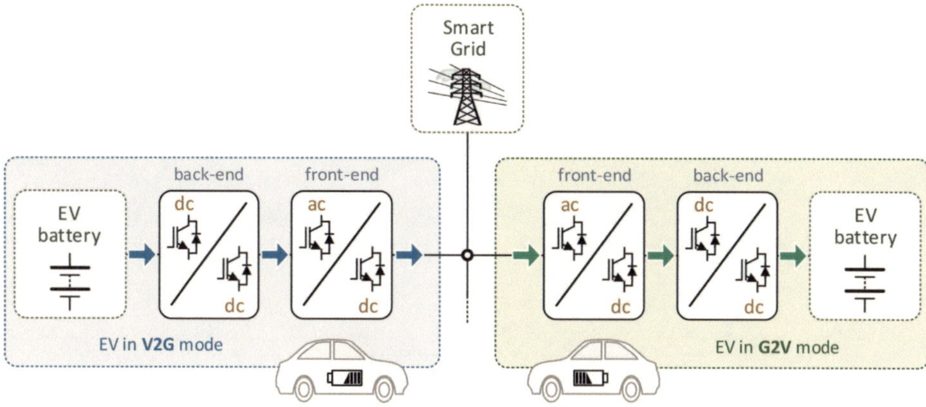

Figure 13.
Power transfer between EV batteries with EVs connected to the same electrical grid, one operating in G2V mode and the other in V2G mode.

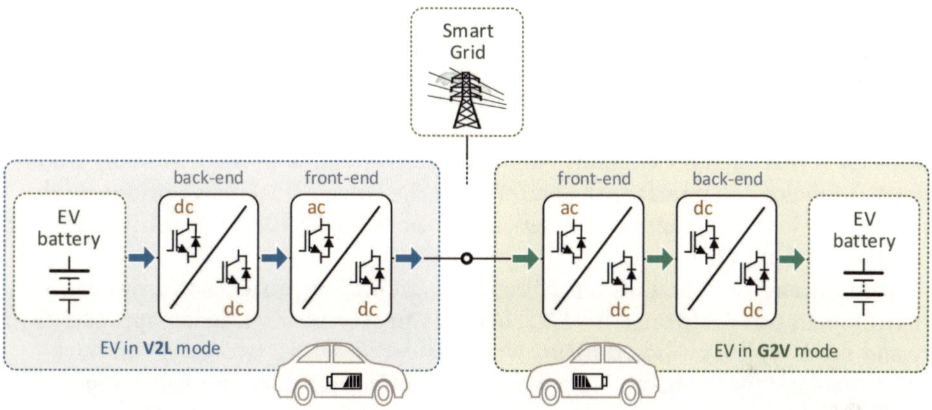

Figure 14.
Power transfer between EVs using both front-end and back-end power stages of both EVs, and without using electrical power grid interface.

in the G2V mode. The possibility of the EV operation as a voltage source is presented in more detail in Section 7 and is identified as vehicle-to-load (V2L). Contrarily to the previous case, this approach of V2V does not require the interface with the electrical grid, and the front-end power stage of the EV that operates as a voltage source is controlled by voltage feedback (operation that emulates the electrical grid). **Figure 14** illustrates this case of power transfer between EVs using both front-end and back-end power stages of both EVs and without the electrical grid interface.

5.2 V2V power transfer using the back-end power stages

A direct V2V power transfer without the need for the connection to an electrical grid was proposed and analyzed in [169] and developed in [170], with two EVs being connected by the ac-side of each on-board EVBC. With this approach, it is possible to provide power to an EV that has its batteries completely discharged and, therefore, cannot move to a charging station or to a power outlet to be charged. Despite the use of the ac-side converters, the power transfer is performed in dc. **Figure 15** illustrates this case of power transfer directly between EV batteries only using the back-end power stages. Accordingly, this V2V approach is more efficient than the previously referred combination of V2G and G2V and allows the power

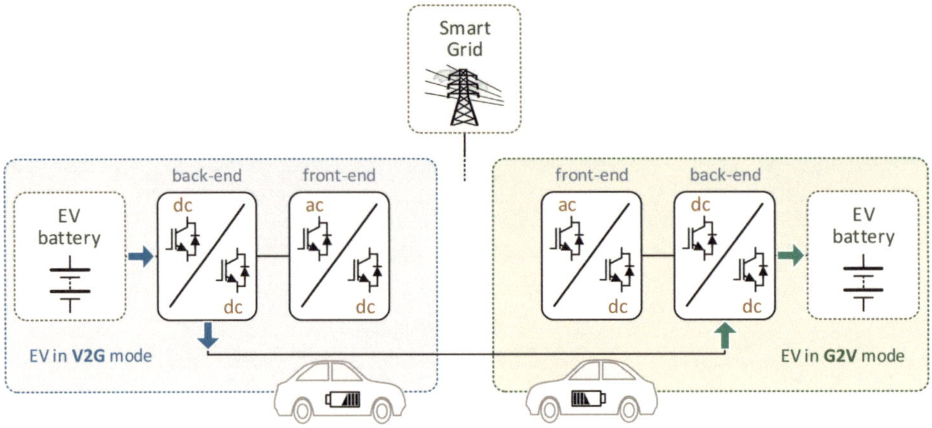

Figure 15.
Power transfer between EVs using only the back-end power stages of both EVs, and without using the electrical power grid interface.

transfer between EVs in remote areas, that is, without the need of the electrical grid. In [171], another topology was presented for V2V power transfer that uses two back-end dc-dc converters in each EV, plus a front-end ac-dc converter, whose dc-links represent the connection point of the two EVs. The main disadvantage of this topology is that each EV contains two dc-dc converters, with the first being a non-isolated topology to interface the battery and the latter a dual active bridge dc-dc converter. Moreover, these converters are on-board, with the ac-dc converter being the only off-board converter.

An efficiency comparison regarding different V2V approaches in a simulation environment was performed in [172], being compared power transfer approaches in ac and dc. A typical on-board EVBC was considered, with a two-quadrant buck-boost topology for the back-end power stage, and a four-quadrant full-bridge topology for the front-end power stage. The connection between the EVs was performed through the dc-links formed by the power stages, with the front-end power stage not being used. This connection formed a split-pi buck-boost converter, resulting in a dc-dc converter capable of operating in four-quadrants. This is advantageous, meaning that a given EV battery can provide power to another, regardless of its voltage being higher or lower than the supplying battery. Moreover, despite the connection forming two dc-dc converters, it is possible to perform a power transfer with only one converter activated. In this way, different control modes for performing the power transfer can be considered, namely by controlling the dc-link voltage or only controlling the battery current for each EV.

Besides the aforementioned wired V2V mode, wireless power transfer regarding V2V operation is also possible for two EVs [171, 173]. Wireless power transfer takes V2V power transfer a step further, allowing not only the power transfer between EVs in remote areas, but also the power transfer between EVs without the need of being stopped. The V2V concept is a relatively recent topic of research and it is expected that new developments would take place in the next few years.

6. Unified technologies for the vehicle electrification

The main purpose of an EV, as well as with any other type of vehicle, is to perform transportation. In order to perform this function in an EV, the electrochemical energy stored in the EV batteries is controlled, via power electronics converters, to drive the electric motor of the EV, which in turn transforms the supplied electrical energy into mechanical energy, making the EV able to move. Additionally, the reverse process, that is, regenerative braking, is also possible, since an electric motor can also behave as an electric generator. It should be noted that the power electronics converters responsible for the EV motor driver should be bidirectional in order to perform regenerative braking. On the other hand, an EV contains also an on-board EVBC, making it possible to charge the EV batteries with power from a domestic power outlet, for instance. Contrarily to an off-board EVBC, which operates with power levels classified as Level 3 (50–100 kW) and, hence, provide fast battery charging, the on-board EVBC are only framed in Level 1 (1.4–1.9 kW) and Level 2 (4–19.2 kW), offering slow battery charging operation [174]. Compared with the power electronics converters used for the EV motor driver, the EVBC has a substantially lower power rating, since the EV motor driver needs to be sized for a power level above (or equal to, in the limit) the electric motor nominal power. As happens with internal combustion engine vehicles, the range of available power values for EV motors is relatively large, ranging from dozens of kW, such as Renault Zoe (65 kW) [175] or the first generation Nissan Leaf (80 kW) [176], to several hundreds of kW, such as Tesla Model S (451 kW for the P100D model) [177]. Power levels of even MW can be also found, as in supercar Rimac C Two (1.048 MW) [178], for example. As it can be seen, even for lower powered EVs, the EV motor driver has a power rating several times higher than the on-board EVBC.

6.1 Integrated battery chargers for the vehicle electrification

Based on the previous analysis, an EV comprises two main groups of power electronics converters: the EV motor driver and the EVBC, with the first being used to perform the EV movement and the latter to supply power to the EV batteries. **Figure 16** illustrates this case. Accordingly, only one group of power electronics converters is used at a time: the EV either is being used for traveling, with the only possibility of charging its batteries being through regenerative braking, or is charging its batteries through the on-board (or an off-board) EVBC, with the EV being stopped in this situation. In both cases, there is no superposition of active groups of power electronics converters, attributing some redundancy to these converters. This redundancy gave rise to the concept of integrated battery chargers, that is, only one single group of power electronics converters is used to perform both the traction (motor driver) and the battery charging operations. Besides reducing the

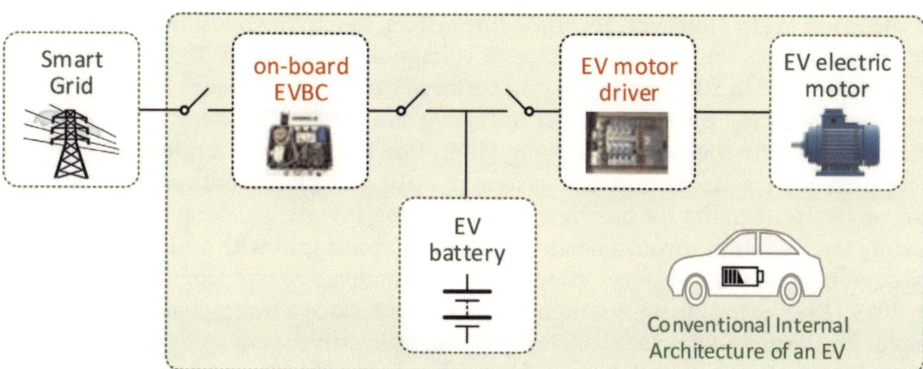

Figure 16.
Conventional internal architecture of an EV constituted by the on-board EVBC and the EV motor driver.

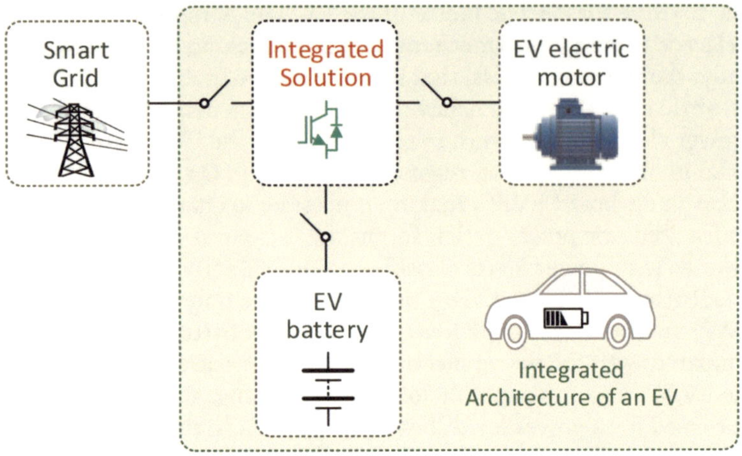

Figure 17.
Integrated architecture of an EV used for two purposes: On-board EVBC and EV motor driver.

required hardware, this approach furnishes the EV with a fast on-board EVBC, since the battery charging power level is established by the EV motor driver. **Figure 17** illustrates an integrated architecture of an EV used for both purposes: on-board EVBC and EV motor driver.

The first publication on integrated battery chargers dates back to 1983 with a USA Department of Energy/NASA report [179], followed by a journal publication of the same author 2 years later [180], when EVs were far from having the popularity of the twenty-first century second decade. In this approach, a 3.6-kW resonant inverter based on silicon-controlled rectifiers (SCRs) was used. A few years later, Rippel and Cocconi filled patents regarding integrated battery chargers [181–183] In the first of these patents [181], dating to 1990, a connection to a single-phase ac electrical grid was available through a diode full-bridge rectifier, with the traction inverter operating as a boost dc-dc converter to charge the EV batteries. Despite an external inductor being used for the boost operation, the authors referred that the leakage inductance of the motor windings could be used instead, although leading to a high ripple in the battery current. In Rippel and Cocconi [182], a scheme was proposed for two induction motors or, alternatively, a motor with two sets of windings. This system comprised two three-phase inverters and used the motor windings as the boost dc-dc converter inductors. Similarly to the previous proposal, this system considered the connection to a single-phase ac electrical grid. In Cocconi [183], the previous work was extended to single-phase and three-phase ac electrical grids. However, for these three cases, due to the boost operation of the traction inverter, the electrical grid peak voltage should be lower than the EV battery voltage. In 2001, an integrated battery charger for an electric scooter was proposed, with the traction inverter operating as a three-phase boost dc-dc converter to perform the battery charging [184]. Power factor correction (PFC) characteristics were added to a similar system in 2010, as well as a bidirectional dc-dc converter between the EV battery and the traction inverter, making it possible to charge the EV battery from a single-phase ac electrical grid with a higher or lower peak voltage than the battery voltage [185]. An innovative topology was proposed in 2013 [186], using an eight switch inverter to interface a three-phase induction motor and a single-phase ac electrical grid. An innovative topology termed as multi-source inverter was recently proposed for plug-in hybrid EVs, aiming to connect multiple dc sources to the same ac output though a single power conversion stage [187, 188].

6.2 Integrated battery chargers: the electric motor perspective

Concerning electric motors, switched reluctance motors have been gaining interest due to their constructive simplicity, low size and weight, and low cost. An integrated battery charger for two-phase switched reluctance motors with connection to a single-phase ac electrical grid was proposed in 2000 [189]. In this case, the integrated battery charger behaved as a flyback dc-dc converter, with an auxiliary coupled winding being used for the battery charging operation. In 2009, a similar system was proposed for three-phase switched reluctance motors, with the traction inverter forming a PFC topology [190]. Two motor windings were used as input filters of the diode bridge rectifier, while the third winding was used as the inductor of the boost dc-dc converter. Two years later, the same authors proposed a modification of this system, adding buck-boost PFC charging functionalities, by changing the traction inverter topology [191]. In both cases, the system was connected to a single-phase ac electrical grid. In 2014, an integrated battery charger for switched reluctance motors applicable to plug-in hybrid EVs was proposed, allowing the battery charging operation from the EV internal combustion engine or ac electrical grids, either single-phase or three-phase [192]. One year later, similar systems were proposed for four-phase switched reluctance machines, with [193] proposing increased functionalities, such as V2G and V2H, and [194] proposing a system based on a dual converter, supporting battery charging from both dc and single-phase ac electrical grids. In 2017, an integrated battery charger based on a four-level converter for a three-phase switched reluctance motor was proposed, for application in plug-in hybrid EVs [195, 196]. In both cases, the batteries could be charged from the internal combustion engine or from a three-phase ac electrical grid.

It is relevant to note that integrated battery chargers encompassing galvanic isolation are also possible. The previously referred system proposed in [189] for switched reluctance motors achieved galvanic isolation through a flyback dc-dc converter, but its battery charging efficiency was low (25%). Two galvanically isolated integrated battery chargers, to be used in industrial EVs, were proposed in 2005 [197]: one of the systems aimed for 1.5 kW dc motor powered pallet trucks, with galvanic isolation being accomplished with a Ćuk converter; the other aimed for 6 kW wound rotor induction motor powered forklift, in which galvanic isolation was accomplished by the motor itself (while the stator windings were connected to the inverter, the rotor windings were connected to a three-phase ac electrical grid). Integrated battery chargers for EVs using a motor/generator set and winding reconfiguration to achieve galvanic isolation were proposed in 2011 [198] and 2013 [199, 200]. Despite adding safety to the battery charging process, galvanically isolated integrated battery chargers are disadvantageous in terms of size, weight, cost, and efficiency compared to non-isolated topologies and, therefore, are less analyzed in the literature than the latter.

The interest for multiple motor powertrains has been increasing, as well as integrated battery chargers for such purpose. In fact, an integrated charger for a four in-wheel motor EV was proposed in 1995 [201]. Four inverters and four sets of three-phase windings were combined to achieve an interleaved operation, with two motors/inverters forming a single-phase ac-dc converter to interface the electrical grid and the other two motors/inverters forming a two-phase bidirectional interleaved buck-boost dc-dc converter to interface the EV battery. In 2015, a dual motor/generator set was proposed as an integrated battery charger to be connected to a single-phase ac electrical grid [202]. Integrated battery chargers based on a single motor and a dual inverter are also common, being proposed in 2015 [203] a topology for charging the secondary battery of EVs through the main battery, with the dual inverter and the motor windings interfacing both batteries. However, this system required an additional on-board EVBC. In 2018 [204], a similar solution capable of charging both batteries simultaneously from a single-phase ac electrical grid was proposed, being necessary to add a diode-bridge rectifier to interface the electrical grid. A more complex solution comprising galvanic isolation was proposed in 2016 [205] for interfacing a three-phase ac electrical grid, using a diode bridge rectifier and a full-bridge inverter per phase to connect with each of the three primary windings of a magnetic combination transformer, with the only secondary winding being connected to a diode bridge rectifier which, in turn, was followed by the EV battery.

Besides multiple motor and multi-inverter topologies, integrated battery chargers based on multi-phase motors are also commonly found in the literature. An integrated battery charger for a powertrain based on a five-phase motor was presented in 2016 [206], which was capable of fast battery charging, that is, interfacing with a three-phase ac electrical grid. Slow [207] and fast [208] battery charging operations concerning integrated battery chargers with five-phase, six-phase, and nine-phase motors were analyzed in the same year by the same authors, and galvanic isolation was considered for six-phase motors the next year [209]. Further reading concerning multi-phase motors and integrated battery chargers for these can be found in [210–212].

7. Vehicle electrification: innovative modes contextualized with smart homes and smart homes

The possible structures that can be implemented for an EVBC were presented in Section 2 and the different technologies targeting the vehicle electrification in Sections 3–6. Using the previous sections as support, this section introduces new opportunities for the smart grids and smart homes arising from the EV flexible operation.

7.1 EV battery charger: on-board

The main operation modes of an on-board EVBC are presented considering the restrictions and also the offered opportunities when integrated in smart grids and smart homes scenarios. **Figure 18** illustrates a smart home with an on-board EVBC plugged-in. As shown, a bidirectional communication is necessary for establishing a power management control between the smart home, the smart grid, the electrical appliances, and the EV. In fact, the power management at the smart home level is used for communicating with the EVBC and with the controlled electrical appliances aiming to define control strategies based on schedules of operation. On the other hand, the power management at the smart home communicates with the power management of the smart grid.

7.1.1 Operation mode: grid-to-vehicle (G2V)

Nowadays, the G2V mode is the existing mode on commercial EVs, which is related with the EV battery charging. **Figure 19** illustrates an on-board EVBC plugged-in at a smart home. As shown, a unidirectional power flow is established with the electrical grid, but a bidirectional communication is established for communicating the charging status and for defining set-points of operation. In G2V, the value of the grid-side current is independent of the other electrical appliances. Since the current is limited by the home switch-breaker, if the consumed current exceeds the nominal value, then the switch-breaker will be triggered. In order to overcome

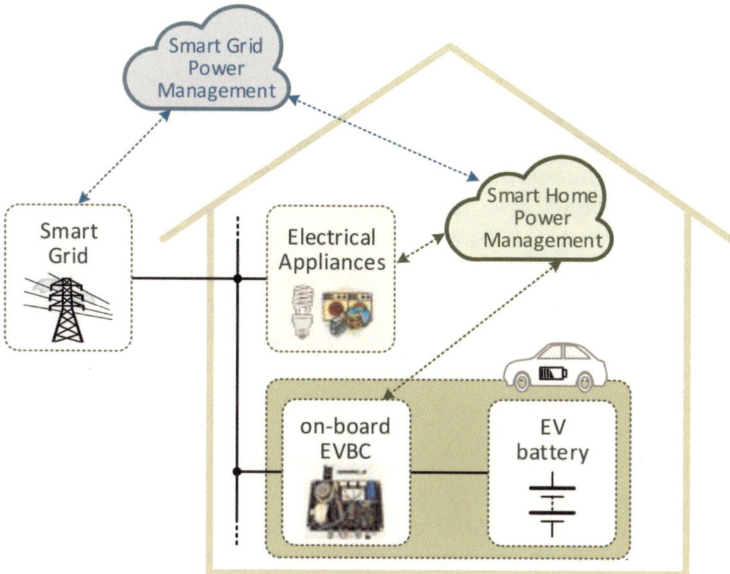

Figure 18.
An on-board EVBC plugged-in at a smart home.

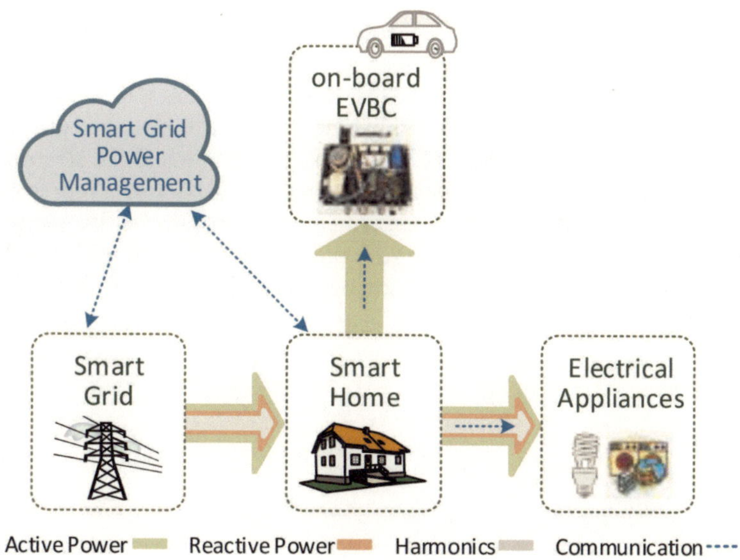

Figure 19.
On-board EVBC: G2V operation mode.

this situation, the smart home power management forces the EVBC to stop the G2V mode, representing a disadvantage of this mode.

Analogously to the aforementioned G2V mode, the flexible G2V mode refers to a situation when the EV charging power is adjusted according to the status of the other electrical appliances [95]. For example, the value of the charging power can be adjusted based on the injected power from RES as a contribution to balance the production/consumption from the smart home perspective. Moreover, it can be performed without harming power quality aspects. Also in this mode, it is fundamental to establish a bidirectional communication between the EVBC and the smart home power management.

7.1.2 Operation mode: vehicle-to-grid (V2G)

The V2G mode denotes a state related with the possibility of a bidirectional operation also in terms of power flow: the EV is used to return part of the stored energy back to the electrical grid. This mode is performed in convenience of the smart grid or smart home power management, as well as in the convenience of the EV user. Therefore, the EV is seen as a flexible ESS allowing a support for the grid stability. Furthermore, this mode entails communication with a smart grid aggregator targeting to outline schedules for the EVBC operation, as well as the quantity of power that must be returned back to the electrical grid. **Figure 20** illustrates this operation mode when the EV is plugged-in at the smart home, permitting a dual opportunity: the flexible operation in V2G mode for the smart home and/or for the smart grid.

7.1.3 Operation mode: vehicle-to-load (V2L) - as voltage source

In the G2V/V2G modes, the controllability of the EVBC is performed, respectively, only in relation to absorb/inject active power. In both G2V/V2G case, a current feedback control is applied. Instead, a new opportunity for the EV operation is associated with the EVBC controllability as a voltage source for supplying electrical appliances (loads). This operation is only valid while the EV is not plugged-in to the electrical grid, which is denominated as V2L (where a voltage feedback control is applied, meaning that the voltage waveform is forced by the EVBC and the current waveform by the electrical appliances) [213]. **Figure 21** illustrates the principle of operation of the V2L mode. The relevance of the V2L mode is linked with the option to use the EV in isolated locations from the electrical grid (for instance, in extreme circumstances of catastrophic events when the electrical grid is unavailable or in campsites). This operation is very applicable and represents a new support offered by the EV, however, since it requires to use the energy stored in the battery, the battery state-of-charge is obligatory managed with the EV owner agreement (for instance, conserving an acceptable state-of-charge for the next travel). Similar opportunity was before recognized by Nissan (the

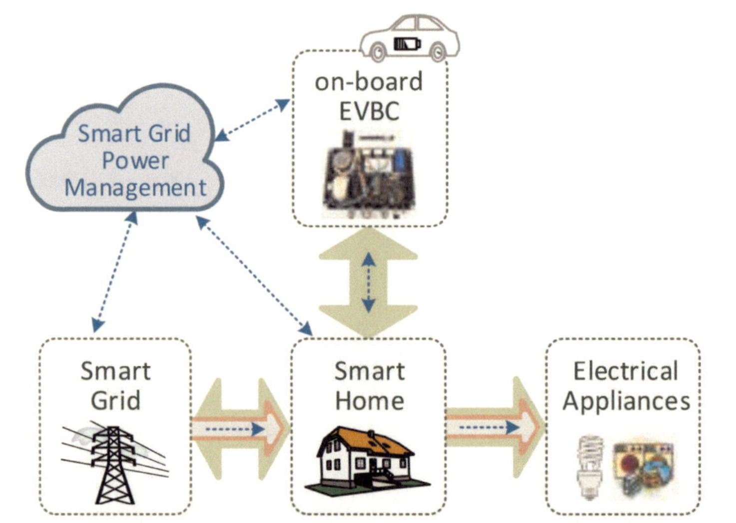

Figure 20.
On-board EVBC: V2G operation mode.

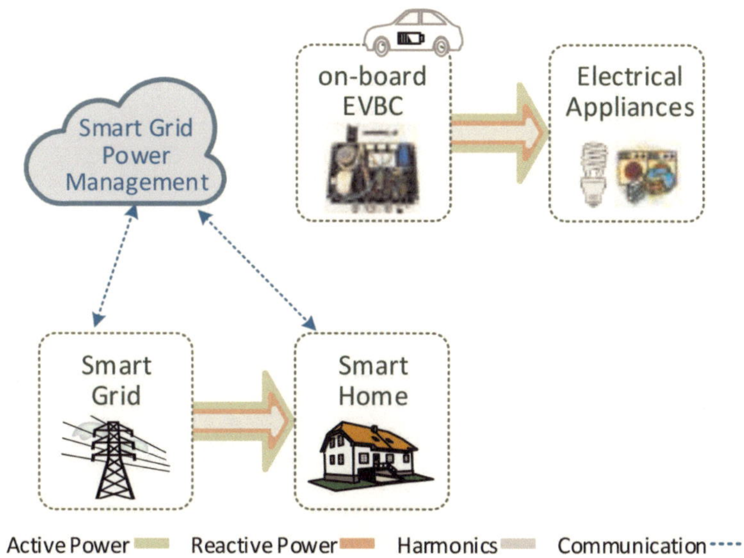

Figure 21.
On-board EVBC: V2L operation mode (as a voltage source).

"LEAF-to-Home" project), but requiring an external "EV Power Station," limiting the application of this concept to the location where the system is permanently installed [214]. Consequently, the presented V2L mode assumes a greater field of application, since it can be used generically with the EV in the place where it is parked.

7.1.4 Operation mode: vehicle-to-home (V2H) - as uninterruptible power supply (UPS)

As a sequence of V2L mode, emerges the possibility of the EVBC operation with features based on an off-line uninterruptible power supply (UPS) [215]. This is especially dedicated for smart homes in the existence of a power failure, where the EVBC starts to operate as a voltage source practically instantaneously. In this mode, it is required a communication from the smart home to the EVBC notifying about the power outage and a communication from the EVBC to the smart home to inform about the battery state-of-charge (for instance, permitting to establish a control based on selecting priority electrical appliances). **Figure 22** illustrates the V2H mode as a UPS contextualized into a smart home, which evidently recognizes the operation disconnected from the electrical grid. As in the previous mode, the grid-side converter (front-end power stage) is controlled with a voltage feedback, however, it is obligatory to measure the electrical grid voltage for noticing the power failure (in this event, the smart home is disconnected from the electrical grid almost instantaneously and the EVBC starts its process). When the voltage is restored, such situation is identified by the EVBC and, subsequently, it starts the synchronization with the phase of the voltage targeting the transition to the normal mode, when the electrical grid supplies power for the smart home. Posteriorly, the EVBC can stay in an idle state or it can return to a G2V/V2G mode.

7.2 EV battery charger: off-board

The foremost opportunities for an off-board EVBC are addressed in this section targeting a contextualization with smart grids. It must be highlighted that the identified opportunities are independent from the off-board EVBC classification as

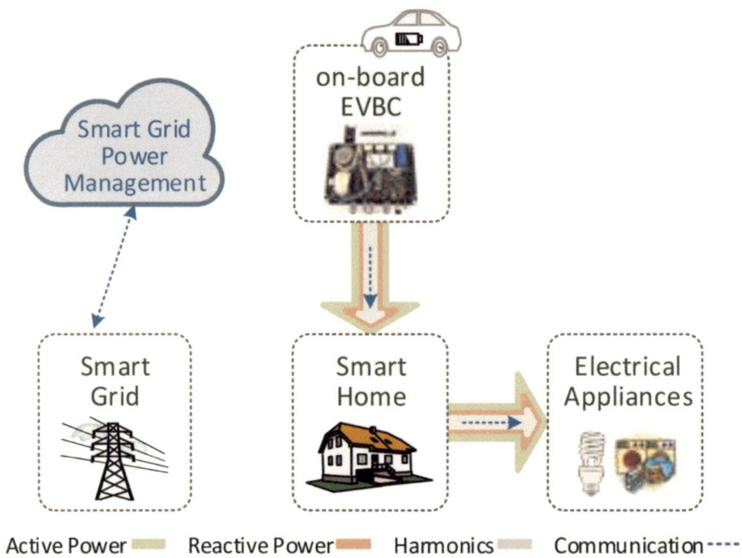

Figure 22.
On-board EVBC: V2H operation mode (as an off-line uninterruptible power supply).

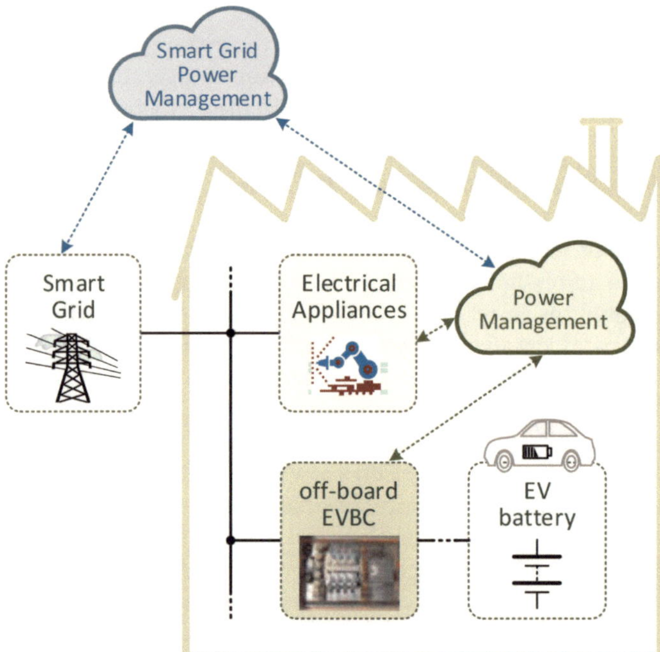

Figure 23.
An off-board EVBC and an EV plugged-in at an industry.

slow, semi-fast, fast, or ultra-fast (in single-phase or three-phase interfaces). **Figure 23** illustrates an off-board EVBC into an industry. Since it is an off-board EVBC, it is always connected to the electrical grid even without any plugged-in EV. Also in an off-board EVBC, it is fundamental a bidirectional communication, allowing to transmit data from the off-board EVBC (for instance, the information of the battery state-of-charger) and transmit set-points of operation to the off-board EVBC. This communication is indispensable for the smart grid contextualization.

7.2.1 Operation mode: grid-to-vehicle and vehicle-to-grid

An off-board EVBC also enables the G2V/V2G modes, but the core variance, when compared with an on-board EVBC, is the operating power, which is significantly higher (the power is higher, but it is used for shorter periods of time). **Figure 24** illustrates an off-board EVBC operating in G2V/V2G modes, where a

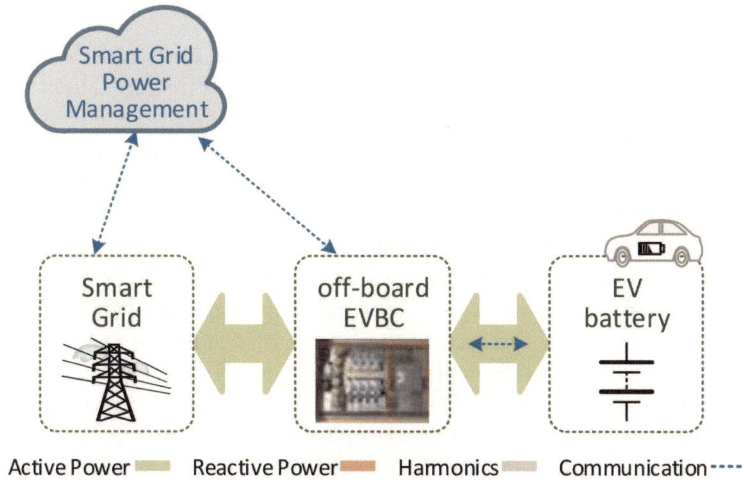

Figure 24.
An off-board EVBC with an EV plugged-in into the electrical power grid: G2V/V2G modes.

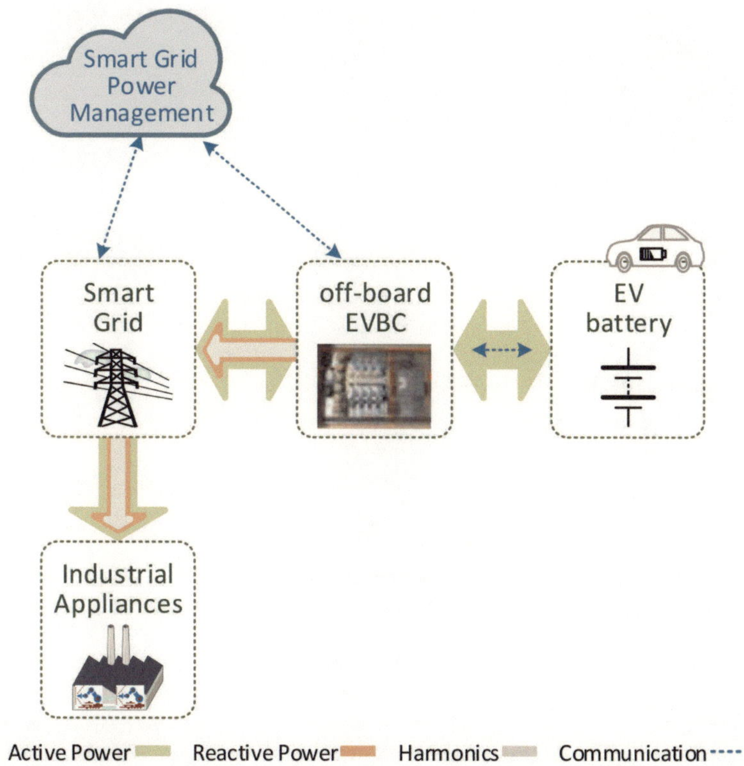

Figure 25.
An off-board EVBC with an EV plugged-in into the electrical power grid: operation as a power quality compensator.

bidirectional power flow and a bidirectional communication is identified. The V2G mode is interesting, but in the case of an off-board EVBC, its use is very particular, because when the EV is parked and plugged-in the goal is to charge the battery as fast as possible, so if the process is interrupted for the V2G mode, the charging will take longer.

7.2.2 Operation mode: power quality compensator

As previously identified, an off-board EVBC can be used in G2V/V2G modes, a situation that occurs as fast as possible (where a high power value in a short period of time is necessary). Therefore, after the EV charging, the off-board EVBCs may be out of operation throughout some periods, that is, until another EV arrives to charge. Accordingly, a new opportunity is recognized for the off-board EVBC when the EV is not plugged-in, which is linked with the support to the smart grid in terms of power quality (mainly, the issues corresponding to low power factor, current imbalances in three-phase systems, and harmonic current). Furthermore, the existing opportunity of this mode is additionally attractive, by the reason that it can be accomplished while the EV plugged-in (for instance, without jeopardizing the off-board EVBC or without using the stored energy in the EV battery, the G2V/V2G modes can also be performed) or without any EV plugged-in. This means that it is not necessary to transfer active power from the electrical grid to the EV or vice-versa. Additionally, it is not required extra hardware for this additional mode of operation related with power quality coverage. **Figure 25** illustrates an off-board EVBC,

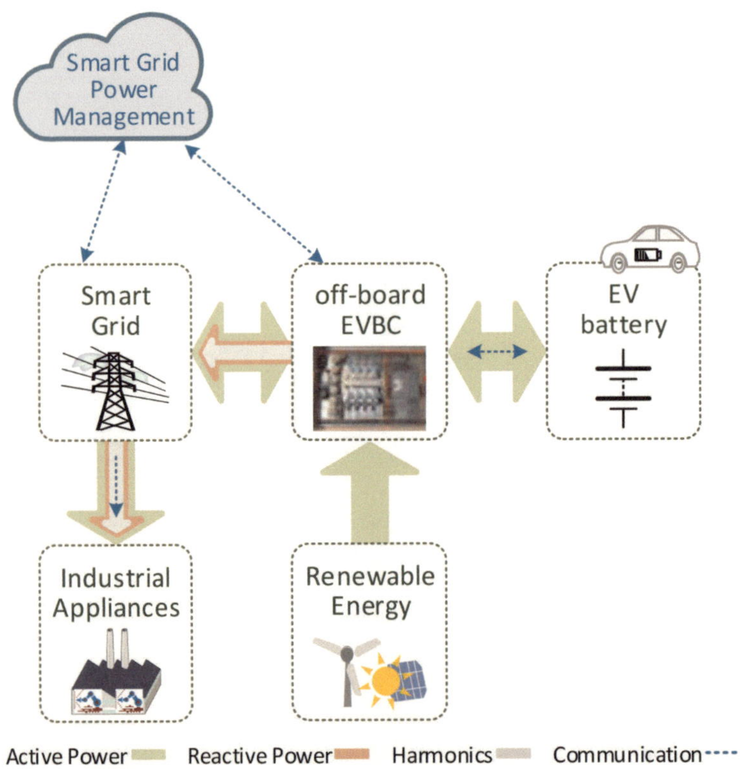

Figure 26.
An off-board EVBC with an EV plugged-in into the electrical power grid: unified operation with RES and as a power quality compensator.

where is highlighted this opportunity. In this case, the power quality problems are determined by the linear and nonlinear electrical appliances within the industry.

7.2.3 Unified operation: power quality compensator and interface of renewable energy sources

The possibility for exchanging power with the electrical grid and for compensating power quality issues was previously presented. Moreover, knowing the stimulus of RES for the progress of smart grids, also to mitigate the impact of the EV required power from the electrical grid, their installation close to the off-board EVBC is of pertinent importance. As the most pertinent example of RES, solar photovoltaic panels can be mounted in EV charging stations, as well as in industries. This is an advantage for solutions as described in this section. Since the off-board EVBC and RES require similar front-end power stages, the identified opportunity consists of unifying both systems targeting a single interface with the electrical grid. **Figure 26** illustrates this opportunity, requiring a common dc-link for both EV and RES. The utmost advantage of this opportunity is about the efficiency. In this condition, as the EV establishes a direct interface with the RES for the charging (over the dc-link and requiring less power stages), it is conceivable to boost the efficiency when compared to customary solutions (based on various front-end and back-end power stages).

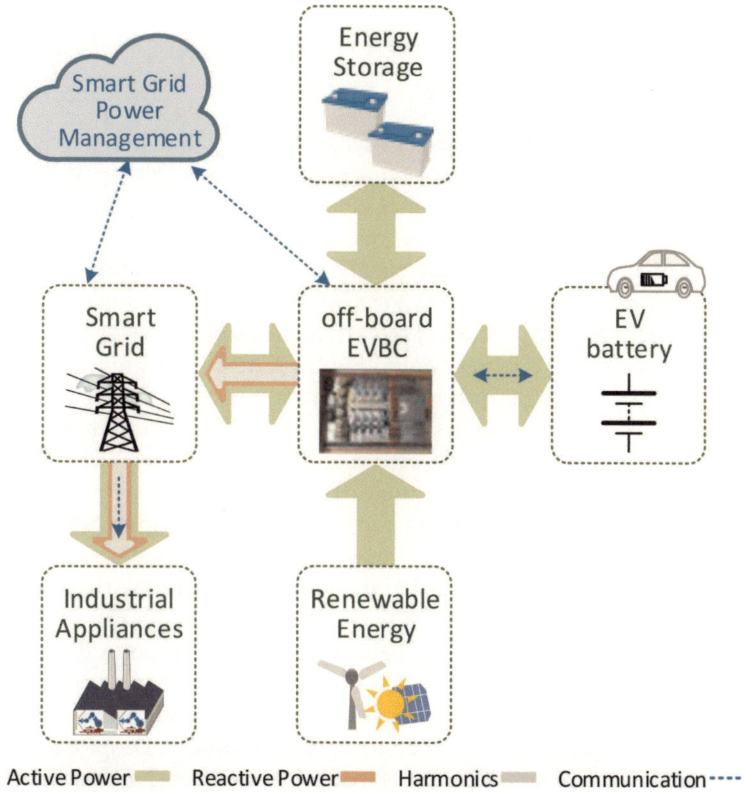

Figure 27.
An off-board EVBC with an EV plugged-in into the electrical power grid: unified operation with RES, ESS, and as a power quality compensator.

7.2.4 Unified operation: power quality compensator and interface of energy storage systems and renewable energy sources

Based on the previously identified opportunity, adding a bidirectional dc interface to the off-board EVBC results in a new opportunity for interfacing an ESS (bidirectional power transfer, charging or discharging, with the dc-link). Consequently, in this circumstance, based on the off-board EVBC, a whole system is offered for the smart grid: interface of G2V/V2G modes; interface of a RES; and interface of a flexible ESS. **Figure 27** illustrates this new opportunity (based on the common dc-link, a single interface with the electrical grid is considered). In this approach, for instance, the power from the RES can be injected directly into the EV (as well as to the ESS), avoiding the electrical grid. Consequently, in this process, fewer power stages are required, allowing to improve the efficiency of the process. A pertinent aspect is also associated when the EV is plugged-in. Since it requires a high value of power in a short period of time, therefore, the ESS is an important influence to avoid power fluctuations in the electrical grid side (in this occasion, the power for the EV can be provided by the ESS). On the other hand, in a situation where the EV is not plugged-in, the RES and the ESS are integrated through the same system, permitting the operation similar to a load shift system (basically, the power production from RES can be stored in the ESS for a posterior use, when convenient for the electrical installation). Moreover, even in a case where the EV is not plugged-in, the RES is not producing, and the ESS is not required, the off-board EVBC can operate for compensating the aforementioned problems of power quality (directly caused by the industrial appliances or in a selective strategy for the smart grid).

8. Conclusions

In this book chapter, technologies, challenges, and a global perspective for the vehicle electrification in smart grids are presented. The new reality of shifting the transportation sector targeting the vehicle electrification, mainly with plug-in electric vehicles (EV), is boosted by climate concerns. However, this new paradigm also promotes a set of emergent technologies, such as: power electronics for on-board and off-board battery charging systems; communication technologies; wireless power transfer for charging processes; bidirectional power transfer in vehicle-to-vehicle mode; unified technologies combining the battery charging system and the motor driver based on a single system; and operation modes of the EV, both on-board and off-board, in smart homes and smart grids. The importance of these emergent technologies for the vehicle electrification is described along this book chapter, as well as the relation among them. The identified EV battery charging operation modes can be performed independently of the charging system structure (i.e., the number and types of power stages for the on-board and off-board charging system). Moreover, since some operation modes only require the front-end power stage (ac-dc converter), technologies of wireless power transfer can also be considered. Similarly, unified technologies of battery charging and motor driver can also be considered for the implementation of the presented operation modes. Furthermore, combined technologies of wireless power transfer and unified systems are also possible in the implementation of some operation modes. Despite the relevance of these technologies in terms of power transfer, communication technologies are absolutely indispensable for defining the operation modes, establishing a bidirectional link for data transfer and power management between the smart grid or smart home, the user, and the EV. This book chapter covers these technologies, demonstrating the relevance of the vehicle electrification, not only as a new paradigm for the transportation sector, but also as a promoter of smart grids.

Conflict of interest

The authors declare no conflict of interest.

Author details

Vitor Monteiro[1]*, Jose A. Afonso[2], Tiago J.C. Sousa[1], Luiz L. Cardoso[1], Jose Gabriel Pinto[1] and Joao L. Afonso[1]

1 ALGORITMI Research Centre, University of Minho, Guimaraes, Portugal

2 CMEMS-UMinho Center, University of Minho, Guimaraes, Portugal

*Address all correspondence to: vmonteiro@dei.uminho.pt

References

[1] EUROSAT. Energy statistics—An overview. In: Eurostat Regional Yearbook, Edition. Aug 2018

[2] Bose BK. Global warming—Energy, environmental pollution, and the impact of power electronics. IEEE Industrial Electronics Magazine. Mar 2010;**4**(1): 6-17

[3] Bozchalui MC, Cañizares CA, Bhattacharya K. Optimal energy management of greenhouses in smart grids. IEEE Transactions on Smart Grid. Mar 2015;**6**(2):827-835

[4] Palencia JC, González TF, Nakata T. Energy use and CO_2 emissions reduction potential in passenger car fleet using zero emission vehicles and lightweight materials. Energy. 2012; **48**(1):548-565

[5] European Commision, Road transport: Reducing CO_2 emissions from vehicles [Online]. Available from: https://ec.europa.eu/ [Accessed: July 24, 2019]

[6] Khaligh A, Li Z. Battery, ultracapacitor, fuel cell, and hybrid energy storage systems for electric, hybrid electric, fuel cell, and plug-in hybrid electric vehicles: State of the art. IEEE Transactions on Vehicular Technology. Jul 2010;**59**(6):2806-2814

[7] Bishop JD, Martin NP, Boies AM. Cost-effectiveness of alternative powertrains for reduced energy use and CO_2 emissions in passenger vehicles. Elsevier Applied Energy. Jul 2014;**124**: 44-61

[8] Ribeiro B, Brito FP, Martins J. A survey on electric/hybrid vehicles. In: Transmission and Driveline, SAE International. 2010

[9] Boulanger AG, Chu AC, Maxx S, Waltz DL. Vehicle electrification: Status and issues. Proceedings of the IEEE. Jun 2011;**99**(6):1116-1138

[10] Wencong S, Rahimi-Eichi H, Zeng W, Chow M-Y. A survey on the electrification of transportation in a smart grid environment. IEEE Transactions on Industrial Electronics. Feb 2012;**8**(1):1-10

[11] Oviedo RM, Fan Z, Gormus S, Kulkarni P. The reign of EVs? An economic analysis from consumer's perspective. IEEE Electrification Magazine. Jun 2014;**2**(2):61-71

[12] Lubes'n'Greases. Perspective on Electric Vehicles. LNG Publishing Company, Inc., Annual Report; 2019

[13] Irle R. E global EV sales for 2018—Final results [Online]. Available from: http://www.ev-volumes.com/country/total-world-plug-in-vehicle-volumes/ [Accessed: Jul 24, 2019]

[14] Raghavan SS, Khaligh A. Electrification potential factor: Energy-based value proposition analysis of plug-In hybrid electric vehicles. IEEE Transactions on Vehicular Technology. Mar 2012;**61**(3):1052-1059

[15] Amin M, Giacomoni AM. Smart grid —Safe, secure, self-healing. IEEE Power Energy Magazine. Jan 2012:33-40

[16] Gungor VC, Sahin D, Kocak T, Ergut S, Buccella C, Cecati C, et al. Smart grid and smart homes—Key players and pilot projects. IEEE Industrial Electronics Magazine. Dec 2012;**6**:18-34

[17] Moslehi K, Kumar R. A reliability perspective of the smart grid. IEEE Transactions on Smart Grid. Jun 2010; **1**(1):57-64

[18] Galus MD, Vaya MG, Krause T, Andersson G. The role of electric

vehicles in smart grids. Wiley Interdisciplinary Reviews—Energy and Environment. Aug 2013;**2**:384-400

[19] Li D, Jayaweera SK. Distributed smart-home decision-making in a hierarchical interactive smart grid architecture. IEEE Transactions on Parallel and Distributed Systems. Jan 2015;**26**(1):75-84

[20] Monteiro V, Afonso JA, Ferreira JC, Afonso JL. Vehicle electrification: New challenges and opportunities for smart grids. Energies. Dec 2018;**12**(1):1-20

[21] Hashmi M, Hänninen S, Mäki K. Survey of smart grid concepts, architectures, and technological demonstrations worldwide. In: IEEE PES Conference on Innovative Smart Grid Technologies Latin America. Oct 2011. pp. 1-7

[22] Dyke KJ, Schofield N, Barnes M. The impact of transport electrification on electrical networks. IEEE Transactions on Industrial Electronics. Dec 2010;**57**(12):3917-3926

[23] Yu X, Cecati C, Dillon T, Simoes MG. The new frontier of smart grids: An industrial electronics perspective. IEEE Industrial Electronics Magazine. Sep 2011;**5**(3):49-63

[24] Vojdani A. Smart integration: The smart grid needs infrastructure that is dynamic and flexible. IEEE Power and Energy Magazine. Dec 2008;**6**(6):71-79

[25] Güngör VC, Sahin D, Kocak T, Ergüt S, Buccella C, Cecati C, et al. Smart grid technologies: Communication technologies and standards. IEEE Transactions on Industrial Informatics. Nov 2011;**7**(4):529-539

[26] Liu N, Chen J, Zhu L, Zhang J, He Y. A key management scheme for secure communications of advanced metering infrastructure in smart grid. IEEE Transactions on Industrial Electronics. Oct 2013;**60**(10):4746-4756

[27] Sakis Meliopoulos AP, Cokkinides G, Huang R, Farantatos E, Choi S, Lee Y, et al. Smart grid technologies for autonomous operation and control. IEEE Transactions on Smart Grid. Mar 2011;**2**(1):1-10

[28] Blaabjerg F, Guerrero JM. Smart grid and renewable energy systems. In: ICEMS International Conference on Electrical Machines and Systems. Aug 2011. pp. 1-10

[29] Yan B, Luh PB, Warner G, Zhang P. Operation and design optimization of microgrids with renewables. IEEE Transactions on Automation Science and Engineering. Apr 2017;**14**(2):573-585

[30] Ackermann T, Carlini EM, Ernst B, Groome F, Orths A, O'Sullivan J, et al. Integrating variable renewables in Europe: Current status and recent extreme events. IEEE Power Energy Magazine. Dec 2015;**13**(6):67-77

[31] Bragard M, Soltau N, Thomas S, Doncker RWD. The balance of renewable sources and user demands in grids: Power electronics for modular battery energy storage systems. IEEE Transactions on Power Electronics. Dec 2010;**25**(12):3049-3056

[32] Beaudin M, Zareipour H, Schellenberglabe A, Rosehart W. Energy storage for mitigating the variability of renewable electricity sources: An updated review. Journal of Energy for Sustainable Development. Dec 2010;**14**(4):302-314

[33] Tushar MHK, Zeineddine AW, Assi C. Demand-side management by regulating charging and discharging of the EV, ESS, and utilizing renewable energy. IEEE Transactions on Industrial Informatics. Jan 2018;**14**(1):117-126

[34] Alvaro C, Brito FP, Martins J, Rodrigues N, Monteiro V, Afonso JL. Assessment of the use of vanadium redox flow batteries for energy storage and fast charging of electric vehicles in gas stations. ELSEVIER Energy. Nov. 2016;**115**(2): 1478-1494

[35] Hernandez JE, Kreikebaum F, Divan D. Flexible electric vehicle (EV) charging to meet renewable portfolio standard (RPS) mandates and minimize green house gas emissions. In: IEEE ECCE Energy Conversion Congress and Exposition, Atlanta USA. Sep 2010. pp. 4270-4277

[36] Alam MR, Reaz MBI, Ali MAM. A review of smart homes—Past, present, and future. IEEE Transactions on Systems, Man, and Cybernetics—Part C: Applications and Reviews. Nov 2012; **42**(6):1190-1203

[37] Erickson LE, Robinson J, Brase G, Cutsor J. Solar Powered Charging Infrastructure For Electric Vehicles: A Sustainable Development. 1st ed. Taylor & Francis, CRC Press; Dec 2017

[38] Robalino DM, Kumar G, Uzoechi LO, Chukwu UC, Mahajan SM. Design of a docking station for solar charged electric and fuel cell vehicles. In: IEEE International Conference on Clean Electrical Power, Capri, Italy. Aug 2009. pp. 655-660

[39] Tushar W, Yuen C, Huang S, Smith DB, Poor HV. Cost minimization of charging stations with photovoltaics: An approach with EV classification. IEEE Transactions on Intelligent Transportation Systems. Jan 2016;**17**(1): 156-169

[40] Saber AY, Venayagamoorthy GK. Plug-in vehicles and renewable energy sources for cost and emission reductions. IEEE Transactions on Industrial Electronics. Apr 2011;**58**(4): 1229-1238

[41] Martinez IJ, Garcia-Villalobos J, Zamora I, Eguia P. Energy management of micro renewable energy source and electric vehicles at home level. Journal of Modern Power Systems and Clean Energy. Nov 2017;**5**(6): 979-990

[42] Monteiro V, Pinto JG, Exposto B, Ferreira JC, Afonso JL. Smart charging management for electric vehicle battery chargers. In: IEEE VPPC Vehicle Power and Propulsion Conference. Oct 2014. pp. 1-5

[43] Peças Lopes JA, Soares FJ, Almeida PM, Moreira da Silva M. Smart charging strategies for electric vehicles: Enhancing grid performance and maximizing the use of variable renewable energy resources. In: EVS24 International Battery, Hybrid and Fuel Cell Electric Vehicle Symposium. May 2009. pp. 1-11

[44] Tsui KM, Chan SC. Demand response optimization for smart home scheduling under real-time pricing. IEEE Transactions on Smart Grid. Dec 2012;**3**(4):1812-1821

[45] Saber AY, Venayagamoorthy GK. Resource scheduling under uncertainty in a smart grid with renewables and plug-in vehicles. IEEE Systems Journal. Mar 2012;**6**(1):103-109

[46] Pedrasa MAA, Spooner TD, MacGill IF. Coordinated scheduling of residential distributed energy resources to optimize smart home energy services. IEEE Transactions on Smart Grid. Sep 2010;**1**(2):134-143

[47] Zhang T, Chen W, Han Z, Cao Z. Charging scheduling of electric vehicles with local renewable energy under uncertain electric vehicle arrival and grid power price. EEE Transactions on Vehicular Technology. Jul 2014;**63, 6**: 2600-2612

[48] Vithayasrichareon P, Mills G, MacGill IF. Impact of electric vehicles

and solar PV on future generation portfolio investment. IEEE Transactions on Sustainable Energy. Jul 2015;**6**(3): 899-908

[49] Monteiro V, Pinto JG, Afonso JL. Experimental validation of a three-port integrated topology to Interface electric vehicles and renewables with the electrical grid. IEEE Transactions on Industrial Informatics. Jun 2018;**14**(6): 2364-2374

[50] Camus C, Esteves J, Farias T. Integration of electric vehícles in the electric utility systems. In: Electric Vehicles—The Benefits and Barriers. IntechOpen, Sep 2011. pp. 135-158. DOI: 10.5772/16587

[51] Cheng L, Chang Y, Wu Q, Lin W, Singh C. Evaluating charging service reliability for plug-In EVs from the distribution network aspect. IEEE Transactions on Sustainable Energy. Oct 2014;**5**(4):1287-1296

[52] Clement-Nyns K, Haesen E, Driesen J. The impact of charging plug-in hybrid electric vehicles on a residential distribution grid. IEEE Transactions on Power Systems. Feb 2010;**25**(1):371-380

[53] Jiang Z, Tian H, Beshir MJ, Vohra S, Mazloomzadeh A. Analysis of electric vehicle charging impact on the electric power grid: Based on smart grid regional demonstration project—Los Angeles. In: IEEE PES Transmission and Distribution Conference and Exposition-Latin America (PES T & D-LA), Morelia. 2016. pp. 1-5

[54] Hua L, Wang J, Zhou C. Adaptive electric vehicle charging coordination on distribution network. IEEE Transactions on Smart Grid. Nov 2014; **5**(6):2666-2675

[55] Veldman E, Verzijlbergh RA. Distribution grid impacts of smart electric vehicle charging from different perspectives. IEEE Transactions on Smart Grid. Jan 2015; **6**(1):333-342

[56] Ustun TS, Zayegh A, Ozansoy C. Electric vehicle potential in Australia: Its impact on smartgrids. IEEE Industrial Electronics Magazine. Dec 2013;**7**(4): 15-25

[57] Joos G, Dubois MR. Integration of PHEVs and EVs: Experience from Canada. In: IEEE PES General Meeting. Jul 2010. pp. 1-5

[58] Song Y, Yang X, Lu Z. Integration of plug-in hybrid and electric vehicles experience from China. In: IEEE PES Power and Energy Society General Meeting. Jul 2010. pp. 1-6

[59] Bertling L, Carlson O, Lundmark S, Steen D. Integration of plug in hybrid electric vehicles and electric vehicles—Experience from Sweden. In: IEEE PES General Meeting. Jul 2010. pp. 1-3

[60] Pecas Lopes JA, Soares F, Pedro M, Almeida R. Integration of electric vehicles in the electric power systems. Proceedings of the IEEE. Jan 2011;**99**(1): 168-183

[61] Depenbrock M. The FBD-method, a generally applicable tool for analyzing power relations. IEEE Transactions on Power Apparatus and Systems. May 1993;**8**(2):381-387

[62] Monteiro V, Melendez AAN, Ferreira JC, Couto C, Afonso JL. Experimental validation of a proposed single-phase five-level active rectifier operating with model predictive current control. In: IEEE IECON Industrial Electronics Conference. Nov 2015. pp. 3939-3944

[63] Garche J, Jossen A. Battery management systems (BMS) for increasing battery life time. In: IEEE Telecommunications Energy Special Conference. May 2000. pp. 81-88

[64] Manenti A, Abba A, Merati A, Savaresi SM, Geraci A. A new BMS architecture based on cell redundancy. IEEE Transactions on Industrial Electronics. Sep 2011;**58**(9):4314-4322

[65] Monteiro V, Ferreira JC, Melendez AAN, Afonso JL. Model predictive control applied to an improved five-level bidirectional converter. IEEE Transactions on Industrial Electronics. Sep 2016;**63**(9): 5879-5890

[66] Monteiro V, Melendez AAN, Afonso JL. Novel single-phase five-level VIENNA-type rectifier with model predictive current control. In: IEEE IECON Industrial Electronics Conference. Nov 2017. pp. 6413-6418

[67] Monteiro V, Nogueiras Melendez AA, Couto C, Afonso JL. Model predictive current control of a proposed single-switch three-level active rectifier applied to EV battery chargers. In: IEEE IECON Industrial Electronics Conference, Florence, Italy. Oct 2016. pp. 1365-1370

[68] Monteiro V, Pinto JG, Exposto B, Monteiro LFC, Couto C, Afonso JL. A novel concept of unidirectional bridgeless combined boost-Buck converter for EV battery chargers. In: IEEE ISIE International Symposium on Industrial Electronics, Rio de Janeiro Brazil. Jun 2015. pp. 210-215

[69] Monteiro V, Ferreira JC, Melendez AAN, Couto C, Afonso JL. Experimental validation of a novel architecture based on a dual-stage converter for off-board fast battery chargers of electric vehicles. IEEE Transactions on Vehicular Technology. Feb 2018;**67**(2): 1000-1011

[70] Monteiro V, Pinto JG, Exposto B, Afonso JL. Comprehensive comparison of a current-source and a voltage-source converter for three-phase EV fast battery chargers. In: CPE International Conference on Compatibility and Power Electronics, Lisboa, Portugal. Jun 2015. pp. 173-178

[71] Pandey A, Singh B, Singh BN, Chandra A, Al-Haddad K, Kothari DP. A review of multilevel power converters. Journal of the Institution of Engineers. Mar 2006;**8**:220-231

[72] Leon JI, Vazquez S, Franquelo LG. Multilevel converters: Control and modulation techniques for their operation and industrial applications. IEEE Proc. Nov 2017;**105**(11):2066-2081

[73] Rodríguez J, Lai J-S, Peng FZ. Multilevel inverters: A survey of topologies, controls, and applications. IEEE Transactions on Industrial Electronics. Aug 2002;**49**(4):724-738

[74] Leite R, Afonso JL, Monteiro V. A novel multilevel bidirectional topology for on-board EV battery chargers in smart grids. Energies. Dec 2018;**11**(12): 1-21

[75] Nussbaumer T, Raggl K, Kolar JW. Design guidelines for interleaved single-phase boost PFC circuits. IEEE Transactions on Industrial Electronics. Jul 2009;**56**(7):2559-2573

[76] Ayele GT. Challenges of multi-channel interleaved bidirectional power converters and their digital solutions. In: University of Nottingham, Erasmus Mundus Master Course in Sustainable Transportation and Electrical Power Systems. Sep 2015

[77] Krismer F, Biela J, Kolar JW. A comparative evaluation of isolated bi-directional DC/DC converters with wide input and output voltage range. In: IEEE Industry Applications Conference. Vol. 1. Oct 2005. pp. 599-606

[78] Du Y, Zhou X, Bai S, Lukic S, Huang A. Review of non-isolated bi-directional DC-DC converters for plug-

in hybrid electric vehicle charge Station application at municipal parking decks. In: IEEE APEC Applied Power Electronics Conference and Exposition, Palm Springs, CA. Feb 2010. pp. 1145-1151

[79] Monteiro V, Goncalves H, Afonso JL. Impact of electric vehicles on power quality in a smart grid context. In: IEEE EPQU International Conference on Electrical Power Quality and Utilisation. Oct 2011. pp. 1-6

[80] Al-Fuqaha A, Guizani M, Mohammadi M, Aledhari M, Ayyash M. Internet of things: A survey on enabling technologies, protocols, and applications. IEEE Communications Surveys and Tutorials. 2015;**17**: 2347-2376 (fourthquarter)

[81] Gubbi J, Buyya R, Marusic S, Palaniswami M. Internet of things (IoT): A vision, architectural elements, and future directions. Future Generation Computer Systems. 2013;**29**: 1645-1660

[82] Bui N, Castellani AP, Casari P, Zorzi M. The internet of energy: A web-enabled smart grid system. IEEE Network. 2012;**26**(4):39-45

[83] Li JQ, Yu FR, Deng G, Luo C, Ming Z, Yan Q. Industrial internet: A survey on the enabling technologies, applications, and challenges. IEEE Communications Surveys and Tutorials. 2019;**19**(3):1504-1526

[84] Da Xu L, He W, Li S. S. "internet of things in industries: A survey,". IEEE Transactions on Industrial Informatics. 2014;**10**:2233-2243

[85] Zanella A, Bui N, Castellani A, Vangelista L, Zorzi M. Internet of things for smart cities. IEEE Internet of Things Journal. 2014;**1**:22-32

[86] Du R, Santi P, Xiao M, Vasilakos AV, Fischione C. The sensable city: A survey on the deployment and management for smart city monitoring. IEEE Communications Surveys and Tutorials. 2019;**21**(2):1533-1560, Secondquarter (early access)

[87] Islam SR, Kwak D, Kabir MH, Hossain M, Kwak KS. The internet of things for health care: A comprehensive survey. IEEE Access. 2015;**3**:678-708

[88] Baker SB, Xiang W, Atkinson I. Internet of things for smart healthcare: Technologies, challenges, and opportunities. IEEE Access. 2017;**5**: 26521-26544

[89] Khan R, Khan SU, Zaheer R, Khan S. Future internet: The internet of things architecture, possible applications and key challenges. In: IEEE International Conference on Frontiers of Information Technology (FIT). 2012. pp. 257-260

[90] Krco S, Pokric B, Carrez F. Designing IoT architecture(s): A European perspective. IEEE World Forum on Internet of Things (WF-IoT). 2014:79-84

[91] Karl H, Willig A. Protocols and Architectures for Wireless Sensor Networks. John Wiley and Sons; Wiley-Interscience, 1st ed. Oct. 2007

[92] Yang Z, Yue Y, Yang Y, Peng Y, Wang X, Liu W. Study and application on the architecture and key technologies for IOT. In: IEEE International Conference on Multimedia Technology (ICMT). 2011. pp. 747-751

[93] Wu M, Lu TJ, Ling FY, Sun J, Du HY. Research on the architecture of internet of things. IEEE International Conference on Advanced Computer Theory and Engineering (ICACTE). 2010;**5**:V5-V484

[94] Ray PP. A survey of IoT cloud platforms. Future Computing and Informatics Journal. 2016;**1**:35-46

[95] Monteiro V, Carmo JP, Pinto JG, Afonso JL. A flexible infrastructure for dynamic power control of electric vehicle battery chargers. IEEE Transactions on Vehicular Technology. Jun 2016;**65**(6):4535-4547

[96] Rowland C, Goodman E, Charlier M, Light A, Lui A. Designing Connected Products: UX for the Consumer Internet of Things. O'Reilly Media, Inc.; 1st ed. May 2015

[97] Tsai CW, Lai CF, Chiang MC, Yang LT. Data mining for internet of things: A survey. IEEE Communications Surveys and Tutorials. 2014;**16**:77-97

[98] Zhu T, Xiao S, Zhang Q, Gu Y, Yi P, Li Y. Emergent technologies in big data sensing: A survey. International Journal of Distributed Sensor Networks. 2015; **11**:1-13

[99] Buratti C, Conti A, Dardari D, Verdone R. An overview on wireless sensor networks technology and evolution. Sensors. 2019;**9**:6869-6896

[100] Siep TM, Gifford IC, Braley RC, Heile RF. Paving the way for personal area network standards: An overview of the IEEE P802. 15 Working Group for Wireless Personal Area Networks. IEEE Personal Communications. 2000;**7**: 37-43

[101] Baronti P, Pillai P, Chook VW, Chessa S, Gotta A, Hu YF. Wireless sensor networks: A survey on the state of the art and the 802.15. 4 and ZigBee standards. Computer Communications. 2007;**30**:1655-1695

[102] Afonso JA, Maio AJF, Simoes R. Performance evaluation of bluetooth low energy for high data rate body area networks. Wireless Personal Communications. 2016;**90**:121-141

[103] IEEE Standard 802.15.4, 2006— Part 15.4: Wireless Medium Access Control (MAC) and Physical Layer (PHY) Specifications for Low-Rate Wireleess Personal Area Networks (WPANs). Sep 2006

[104] Lu N, Cheng N, Zhang N, Shen X, Mark JW. Connected vehicles: Solutions and challenges. IEEE Internet of Things Journal. Aug 2014;**1**(4): 289-299

[105] Bluetooth SIG. Specification of the bluetooth system. In: Master Table of Contents and Compliance Requirements, Version 5.0. Dec 2016

[106] Hiertz G, Denteneer D, Stibor L, Zang Y, Costa XP, Walke B. Bluetooth low energy mesh networks: A survey. Sensors. 2017;**17**(7):1-19

[107] Silva RB, Afonso JA, Afonso JL. Development and test of an intra-vehicular network based on bluetooth low energy. World Congress on Engineering. Jul 2017;**I**:1-5

[108] Hiertz GR, Denteneer D, Stibor L, Zang Y, Costa XP, Walke B. The IEEE 802.11 universe. IEEE Communications Magazine. 2010;**48**:62-70

[109] Arena F, Pau G. An overview of vehicular communications. Future Internet. 2019;**11**(2):27

[110] Law D, Dove D, D'Ambrosia J, Hajduczenia M, Laubach M, Carlson S. Evolution of Ethernet standards in the IEEE 802.3 working group. IEEE Communications Magazine. 2013;**51**(8): 88-96

[111] Christensen K, Reviriego P, Nordman B, Bennett M, Mostowfi M, Maestro JA. IEEE 802.3 az: The road to energy efficient ethernet. IEEE Communications Magazine. 2010; **48**(11):50-56

[112] Majumder A. Power line communications. IEEE Potentials. 2004; **23**:4-8

[113] Yonge L, Abad J, Afkhamie K, Guerrieri L, Katar S, Lioe H, et al. An overview of the HomePlug AV2 technology. Journal of Electrical and Computer Engineering. 2013;**2013**:1-20

[114] Raza U, Kulkarni P, Sooriyabandara M. Low power wide area networks: An overview. IEEE Communications Surveys and Tutorials. 2017;**19**:855-873

[115] Augustin A, Yi J, Clausen T, Townsley WM. A study of LoRa: Long range and low power networks for the internet of things. Sensors. 2016;**16**:1466

[116] Vejlgaard B, Lauridsen M, Nguyen H, Kovács IZ, Mogensen P, Sorensen M. Coverage and capacity analysis of sigfox, lora, gprs, and nb-iot. In: IEEE Vehicular Technology Conference (VTC Spring). 2017. pp. 4-7

[117] Mangalvedhe N, Ratasuk R, Ghosh A. NB-IoT deployment study for low power wide area cellular IoT. In: IEEE Annual International Symposium on Personal, Indoor, and Mobile Radio Communications (PIMRC). 2016. pp. 1-6

[118] Ouya A, De Aragon BM, Bouette C, Habault G, Montavont N, Papadopoulos GZ. An efficient electric vehicle charging architecture based on LoRa communication. In: IEEE International Conference on Smart Grid Communications (SmartGridComm). Oct 2017. pp. 381-386

[119] Tanembaum AS. Computer Networks. 5th ed. Pearson; Jan 2012

[120] Naik N. Choice of effective messaging protocols for IoT systems: MQTT, CoAP, AMQP and HTTP. In: IEEE International Systems Engineering Symposium (ISSE). 2017. pp. 1-7

[121] IEEE Standard 802.11, 2006. IEEE Standard for Information Technology—Telecommunications and Information Exchange Between Systems Local and Metropolitan area Networks—Specific Requirements—Part 11: Wireless LAN Medium Access Control (MAC) and Physical Layer (PHY) Specifications

[122] D. Culler, S. Chakrabarti, "6LoWPAN: Incorporating IEEE 802.15. 4 into the IP Architecture," (White paper), 2009.

[123] Jebroni Z, Afonso JA, Tidhaf B. Home energy monitoring system towards smart control of energy consumption. In: International Conference on Green Energy and Networking. Cham: Springer; Nov 2018. pp. 40-53

[124] Sousa RA, Monteiro V, Ferreira JC, Nogueiras Melendez AA, Afonso JL, Afonso JA. Development of an IoT system with smart charging current control for electric vehicles. In: IECON Annual Conference of the IEEE Industrial Electronics Society, Washington, DC. 2018. pp. 4662-4667

[125] Ustun TS, Ozansoy CR, Zayegh A. Implementing vehicle-to-grid (V2G) technology with IEC 61850-7-420. IEEE Transactions on Smart Grid. Jun 2013; **4**(2):1180-1187

[126] Nsonga P, Hussain SMS, Garba A, Ustun TS, Ali I. Performance evaluation of electric vehicle ad-hoc network technologies for charging management. In: IEEE PES Asia-Pacific Power and Energy Engineering Conference (APPEEC), Bangalore. 2017. pp. 1-5

[127] Etemad K. Overview of mobile WiMAX technology and evolution. IEEE Communications Magazine. 2008; **46**(10):31-40

[128] Hussain SMS, Ustun TS, Nsonga P, Ali I. IEEE 1609 WAVE and IEC 61850 standard communication based integrated EV charging management in smart grids. IEEE Transactions on

Vehicular Technology. Aug 2018;**67**(8): 7690-7697

[129] Aftab MA, Hussain SMS, Ali I, Ustun TS. IEC 61850 and XMPP communication based energy management in microgrids considering electric vehicles. IEEE Access. Jul 2018; **6**:35657-35668

[130] Ustun TS, Hussain SMS, Kikusato H. IEC 61850-based communication modeling of EV charge-discharge management for maximum PV generation. IEEE Access. Jan 2019;**7**: 4219-4231

[131] Böttigheimer M, Maier D, Parspour N, Noeren J, Walter R. Validation of the design of an 11 kW inductive charging prototype on a new test bench for WPT-systems. In: IEEE Wireless Power Transfer Conference (WPTC), Montreal, Canada. 2018. pp. 1-4

[132] Shi ZH, Zhi HK, Chong L. Design considerations of 10 kW wireless charger for EV. In: International Conference on Power Electronics Systems and Applications—Smart Mobility, Power Transfer & Security (PESA), Hong Kong. 2017. pp. 1-4

[133] Cirimele V, Diana M, Freschi F, Mitolo M. Inductive power transfer for automotive applications: State-of-the-art and future trends. IEEE Transactions on Industry Applications. Sep 2018; **54**(5):4069-4079

[134] Mohamed AAS, de Almeida FGN, Mohammed O. Harmonics-based steady-state mathematical model of bidirectional inductive wireless power transfer system in V2G applications. In: IEEE Transportation Electrification Conference and Expo (ITEC), Dearborn, MI. 2016. pp. 1-6

[135] Li S, Mi CC. Wireless power transfer for electric vehicle applications. IEEE Journal of Emerging and Selected Topics in Power Electronics. 2015;**3**:4-17

[136] Musavi F, Edington M, Eberle W. Wireless power transfer: A survey of EV battery charging technologies. In: IEEE ECCE Energy Conversion Congress and Exposition, Raleigh, USA. Sep 2012. pp. 1804-1810

[137] Wang S, Dorrell D. Review of wireless charging coupler for electric vehicles. In: IEEE IECON Annual Conference of the Industrial Electronics Society, Vienna, Austria. Nov 2013. pp. 7272-7277

[138] Laporte S, Coquery G, Revilloud M, Deniau V. Experimental performance assessment of a dynamic wireless power transfer system for future EV in real driving conditions. In: ACM International Conference on Future Energy Systems. Jun 2018. pp. 570-578

[139] Ampère A-M. Théorie Mathématique des Phénomènes Électro-Dynamiques, Uniquement Déduite de L'Expérience. 2me, 1883 edn. Paris: A. Hermmann, Librarie Scientifique; 1826

[140] Faraday M. Experimental Researches in Electricity. 2nd, 1849 ed. Vol. 1. London: Richard and John E. Taylor; 1839

[141] Maxwell JC. A Treatise on Electricity and Magnetism. 3rd, 1891 ed. New York: Dover Publications, Inc; 1873

[142] Rautio JC. The long road to Maxwell's equations. IEEE Spectrum. Dec 2014;**51**(12):36-56

[143] Zhang W, Mi CC. Compensation topologies of high-power wireless power transfer systems. IEEE Transactions on Vehicular Technology. Jun 2016;**65**(6): 4768-4778

[144] Yan Z, Zhang Y, Song B, Zhang K, Kan T, Mi C. An LCC-P compensated

wireless power transfer system with a constant current output and reduced receiver size. Energies. 2019;**12**(172): 1-14

[145] Society of Automobile Engineers (SAE) SAE Technical Information Report (TIR) J2954—'Wireless Power Transfer for Light-Duty Plug-in/Electric Vehicles and Alignment Methodology 2016.

[146] Davies RA. Wireless' autos: A Russian dream. Maclean's Magazine. July 1945;**58**(13): 34 and 36

[147] Babat GI. Patent GB657035A. High frequency electric transport system with contactless transmission of energy. 1946;**657**:035

[148] Systems Control Technology Inc. Roadway Powered Electric Vehicle Project Track Construction and Testing Program Phase 3D. California: Palo Alto; 1994

[149] Lashkari K, Shladover SE, Lechner EH. Inductive Power Transfer to an Electric Vehicle. In: Proceedings of 8th International Electric Vehicle Symposium. 1986

[150] Lechner EH, Shladover SE. The roadway powered electric vehicle—An all-electric hybrid system. In: Proceedings of the 8th International Electric Vehicle Symposium. 1986

[151] Boys JT, Covic GA. The inductive power transfer story at the University of Auckland. IEEE Circuits and Systems Magazine. 2015;**15**:6-27

[152] Zhang Z, Chau KT. Homogeneous wireless power transfer for move-and-charge. IEEE Transactions on Power Electronics. Nov 2015;**30**(11): 6213-6220

[153] Fujita T, Yasuda T, Akagi H. A dynamic wireless power transfer system applicable to a stationary system. IEEE Transactions on Industry Applications. Jul 2017;**53**(4):3748-3757

[154] Sun Y, Liao Z-J, Ye Z-H, Tang C-s, Wang P-Y. Determining the maximum power transfer points for MC-WPT systems with arbitrary number of coils. IEEE Transactions on Power Electronics. Nov 2018;**33**(11):9734-9743

[155] Sharaf R, Abdel-Rahman AB, Abd El-Hameed AS, Barakat A, Hekal S, Allam A. A new compact dual-band wireless power transfer system using interlaced resonators. IEEE Microwave and Wireless Components Letters. Jul 2019;**29**(7):498-500

[156] Tang Y, Chen Y, Madawala UK, Thrimawithana DJ, Ma H. A new controller for bidirectional wireless power transfer systems. IEEE Transactions on Power Electronics. Oct 2018;**33**(10):9076-9087

[157] Dai X, Li X, Li Y, Hu AP. Maximum efficiency tracking for wireless power transfer systems with dynamic coupling coefficient estimation. IEEE Transactions on Power Electronics. Jun 2018;**33**(6): 5005-5015

[158] Zhong W, Hui SY. Reconfigurable wireless power transfer systems with high energy efficiency over wide load range. IEEE Transactions on Power Electronics. Jul 2018;**33**(7): 6379-6390

[159] Gil A, Sauras-Perez P, Taiber J. Communication requirements for dynamic wireless power transfer for battery electric vehicles. In: IEEE International Electric Vehicle Conference (IEVC), Florence. 2014. pp. 1-7

[160] Rana MM, Xiang W, Wang E, Li X, Choi BJ. Internet of things infrastructure for wireless power transfer systems. IEEE Access. 2018;**6**: 19295-19303

[161] Covic G, Boys J. Modern trends in inductive power transfer for transportation applications. IEEE Journal of Emerging and Selected Topics in Power Electronics. Mar 2013;**1**(1): 28-41

[162] Lu F, Zhang H, Mi C. A review on the recent development of capacitive wireless power transfer technology. Energies. Nov 2017;**10**(11):1-30

[163] Ohira T. A battery-less electric roadway vehicle runs for the first time in the world. In: IEEE MTT-S International Conference on Microwaves for Intelligent Mobility (ICMIM), Nagoya. 2017. pp. 75-78

[164] Liu C, Chau KT, Wu D, Gao S. Opportunities and challenges of vehicle-to-home, vehicle-to-vehicle, and vehicle-to-grid technologies. Proceedings of the IEEE. Nov 2013; **101**(11):2409-2427

[165] Alvaro R, Gonzalez J, Gamallo C, Fraile-Ardanuy J, Knapen L. Vehicle to vehicle energy exchange in smart grid applications. In: IEEE Conference on Connected Vehicles and Expo. 2014. pp. 1-7

[166] You P, Yang Z. Efficient optimal scheduling of charging station with multiple electric vehicles via V2V. In: IEEE International Conference on Smart Grid Communications (SmartGridComm). 2014. pp. 716-721

[167] Alvaro-Hermana R, Fraile-Ardanuy J, Zufiria PJ, Knapen L, Janssens D. Peer to peer energy trading with electric vehicles. IEEE Intelligent Transportation Systems Magazine. 2016;**8**(3):33-44

[168] Masrur MA, Skowronska AG, Hancock J, Kolhoff SW, McGrew DZ, Vandiver JC, et al. Military-based vehicle-to-grid and vehicle-to-vehicle microgrid—System architecture and implementation. IEEE Transactions on Transportation Electrification. Mar 2018;**4**(1):157-171

[169] Li G, Boukhatem L, Zhao L, Wu J. Direct vehicle-to-vehicle charging strategy in vehicular ad-hoc networks. In: IFIP International Conference on New Technologies, Mobility and Security (NTMS). 2018. p. 5

[170] Nasr M, Gupta K, da Silva C, Amon H, Trescases O. SiC based on-board EV power-hub with high-efficiency DC transfer mode through AC port for vehicle-to-vehicle charging. In: IEEE APEC Applied Power Electronics Conference and Exposition. Vol. 1. 2018. pp. 3398-3404

[171] Taghizadeh S, Jamborsalamati P, Hossain MJ, Lu J. Design and implementation of an advanced vehicle-to-vehicle (V2V) power transfer operation using communications. In: IEEE International Conference on Environment and Electrical Engineering and IEEE Industrial and Commercial Power Systems Europe. 2018. pp. 1-6

[172] Sousa TJC, Monteiro V, Fernandes JCA, Couto C, Melendez AAN, Afonso JL. New perspectives for vehicle-to-vehicle (V2V) power transfer. In: IEEE IECON Annual Conference of the IEEE Industrial Electronics Society. 2018. pp. 5183-5188

[173] Mou X. Vehicle-to-vehicle charging system fundamental and design comparison. In: IEEE International Conference on Industrial Technology (ICIT). 2019. pp. 1628-1633

[174] Yilmaz M, Krein PT. Review of battery charger topologies, charging power levels, and infrastructure for plug-In electric and hybrid vehicles. IEEE Transactions on Power Electronics. May 2013;**28**(5): 2151-2169

[175] Renault. Renault Zoe Technical Specifications, Imprensa Renault [Online]. Available from: http://imprensa.renault.com.br/upload/produto/fichatecnica/59aecd0b42a81ebb05f9753c7cdf7f31.pdf [Accessed: Jul 17, 2019]

[176] Auto123.com. Technical Specifications: 2013 Nissan Leaf SL [Online]. Available from: https://www.auto123.com/en/new-cars/technical-specs/nissan/leaf/2013/base/sl/ [Accessed: Jul 17, 2019]

[177] Electric Vehicle Database. Tesla Model S P100D [Online]. Available from: https://ev-database.org/car/1075/Tesla-Model-S-P100D. 2019

[178] Automobili R. Rimac Automobili C_Two Hypercar—A Car Alive with Technology [Online]. Available from: https://www.rimac-automobili.com/en/hypercars/c_two/ [Accessed: Jul 17, 2019]

[179] Thlmmesch D. Integral Inverter/Battery Charger for Use in Electric Vehicles. USA Department of Energy/NASA; Handbook Department, United States. 1983

[180] Thimmesch D. An SCR inverter with an integral battery charger for electric vehicles. IEEE Transactions on Industry Applications. Jul 1985;**IA-21**(4):1023-1029

[181] Rippel W. Integrated traction inverter and battery charger apparatus. 1990. US4920475A

[182] Rippel W and Cocconi A. Integrated motor drive and recharge system. 1992. US5099186A

[183] Cocconi A. Combined motor drive and battery charger system. 1994. US5341075A

[184] Solero L. Nonconventional on-board charger for electric vehicle propulsion batteries. IEEE Transactions on Vehicular Technology. 2001;**50**(1):144-149

[185] Pellegrino G, Armando E, Guglielmi P. An integral battery charger with power factor correction for electric scooter. IEEE Transactions on Power Electronics. Mar 2010;**25**(3):751-759

[186] Hegazy O, Barrero R, Van Mierlo J, Lataire P, Omar N, Coosemans T. An advanced power electronics interface for electric vehicles applications. IEEE Transactions on Power Electronics. Dec 2013;**28**(12):5508-5521

[187] Dorn-Gomba L, Magne P, Danen B, Emadi A. On the concept of the multi-source inverter for hybrid electric vehicle powertrains. IEEE Transactions on Power Electronics. Sep 2018;**33**(9):7376-7386

[188] Dorn-Gomba L, Guo J, Emadi A. Multi-source inverter for power-split hybrid electric powertrains. IEEE Transactions on Vehicular Technology. Jul 2019;**68**(7):6481-6494

[189] Pollock C, Thong WK. Low-cost battery-powered switched reluctance drives with integral battery-charging capability. IEEE Transactions on Industry Applications. 2000;**36**(6):1676-1681

[190] Chang HC, Liaw CM. Development of a compact switched-reluctance motor drive for EV propulsion with voltage-boosting and PFC charging capabilities. IEEE Transactions on Vehicular Technology. 2009;**58**(7):3198-3215

[191] Chang H, Liaw C. An integrated driving/charging switched reluctance motor drive using three-phase power module. IEEE Transactions on Industrial Electronics. May 2011;**58**(5):1763-1775

[192] Hu Y, Song X, Cao W, Ji B. New SR drive with integrated charging capacity

for plug-in hybrid electric vehicles (PHEVs). IEEE Transactions on Industrial Electronics. Oct 2014;**61**(10): 5722-5731

[193] Hu KW, Yi PH, Liaw CM. An EV SRM drive powered by battery/ supercapacitor with G2V and V2H/V2G capabilities. IEEE Transactions on Industrial Electronics. 2015;**62**(8): 4714-4727

[194] Hu Y, Gan C, Cao W, Li C, Finney S. Split converter-fed SRM drive for flexible charging in EV/HEV applications. IEEE Transactions on Industrial Electronics. Oct 2015;**62**(10): 6085-6095

[195] Gan C, Wu J, Hu Y, Yang S, Cao W, Guerrero JM. New integrated multilevel converter for switched reluctance motor drives in plug-in hybrid electric vehicles with flexible energy conversion. IEEE Transactions on Power Electronics. May 2017;**32**(5): 3754-3766

[196] Ma M, Chang Z, Hu Y, Li F, Gan C, Cao W. An integrated switched reluctance motor drive topology with voltage-boosting and on-board charging capabilities for plug-in hybrid electric vehicles (PHEVs). IEEE Access. 2017;**6**: 1550-1559

[197] Lacressonniere F, Cassoret B. Converter used as a battery charger and a motor speed controller in an industrial truck. In: European Conference on Power Electronics and Applications. Vol. 9. 2005. p. 7

[198] Haghbin S, Lundmark S, Alakula M, Carlson O. An isolated high-power integrated charger in electrified-vehicle applications. IEEE Transactions on Vehicular Technology. Nov 2011; **60**(9):4115-4126

[199] Haghbin S, Lundmark S, Alakula M, Carlson O. Grid-connected integrated battery chargers in vehicle applications: Review and new solution. IEEE Transactions on Industrial Electronics. Feb 2013;**60**(2):459-473

[200] Haghbin S, Khan K, Zhao S, Alakula M, Lundmark S, Carlson O. An integrated 20-kW motor drive and isolated battery charger for plug-in vehicles. IEEE Transactions on Power Electronics. Aug 2013;**28**(8): 4013-4029

[201] Sul S-K, Lee S-J. An integral battery charger for four-wheel drive electric vehicle. IEEE Transactions on Industry Applications. 1995;**31**(5):1096-1099

[202] Woo D-G, Joo D-M, Lee B-K. On the feasibility of integrated battery charger utilizing traction motor and inverter in plug-In hybrid electric vehicles. IEEE Transactions on Power Electronics. Dec 2015;**30**(12):7270-7281

[203] Hong J, Lee H, Nam K. Charging method for the secondary battery in dual-inverter drive systems for electric vehicles. IEEE Transactions on Power Electronics. 2015;**30**(2):909-921

[204] Semsar S, Soong T, Lehn PW. Integrated single-phase electric vehicle charging using a dual-inverter drive. In: IEEE Transportation and Electrification Conference and Expo, ITEC 2018. 2018. pp. 320-325

[205] Li C, Huang W, Cao R, Bu F, Fan C. An integrated topology of charger and drive for electric buses. IEEE Transactions on Vehicular Technology. 2016;**65**(6):4471-4479

[206] Subotic I, Bodo N, Levi E. An EV drive-train with integrated fast charging capability. IEEE Transactions on Power Electronics. 2016;**31**(2):1461-1471

[207] Subotic I, Bodo N, Levi E. Single-phase on-board integrated battery chargers for EVs based on multiphase machines. IEEE Transactions on Power Electronics. 2016;**31**(9):6511-6523

[208] Katic V, Subotic I, Bodo N, Levi E, Dumnic B, Milicevic D. Overview of fast on-board integrated battery chargers for electric vehicles based on multiphase machines and power electronics. IET Electric Power Applications. 2016;**10**(3): 217-229

[209] Subotic I, Bodo N, Levi E, Jones M, Levi V. Isolated chargers for EVs incorporating six-phase machines. IEEE Transactions on Industrial Electronics. Jan 2016;**63**(1):653-664

[210] Levi E. Advances in converter control and innovative exploitation of additional degrees of freedom for multiphase machines. IEEE Transactions on Industrial Electronics. Jan 2016;**63**(1):433-448

[211] Bodo N, Levi E, Subotic I, Espina J, Empringham L, Johnson CM. Efficiency evaluation of fully integrated on-board EV battery chargers with nine-phase machines. IEEE Transactions on Energy Conversion. Mar 2017;**32**(1):257-266

[212] Subotic I, Bodo N, Levi E. Integration of six-phase EV drivetrains into battery charging process with direct grid connection. IEEE Transactions on Energy Conversion. Sep 2017;**32**(3): 1012-1022

[213] Pinto JG, Monteiro V, Goncalves H, Exposto B, Pedrosa D, Couto C, et al. Bidirectional battery charger with grid-to-vehicle, vehicle-to-grid and vehicle-to-home technologies. In: IEEE IECON Industrial Electronics Conference, Vienna, Austria. Nov 2013. pp. 5934-5939

[214] Green Car Congress. Nissan to launch the 'LEAF to Home' V2H power supply system with Nichicon 'EV Power Station' in June [Online] Available from: http://www.greencarcongress.com/2012/05/leafvsh-20120530.html [Accessed May 30, 2012]

[215] Monteiro V, Exposto B, Ferreira JC, Afonso JL. Improved vehicle-to-home (iV2H) operation mode: Experimental analysis of the electric vehicle as off-line UPS. IEEE Transactions on Smart Grid. Nov 2017;**8**(6):2702-2711

Section 4

Industrial Applications

Chapter 8

The Innovative Gaildorf Wind-Water Project Guarantees Reliability of Power Supply

Grażyna Frydrychowicz-Jastrzębska

Abstract

This chapter presents a pilot project, which is an innovative solution related to renewable energy sources (RES). It refers to the integrated system that covers a wind farm (4×3, 4 MW) and a pumped hydro storage PHS (16 MW). Environmental conditions and components of the system were characterised in structural and operational terms. The wind turbines that are part of the system are of considerable height (one of them is even the highest turbine in the world). This is partly the result of the hybrid construction of their towers. Around the bases, there are Bains, which function as a short-term energy storage with the total capacity of 160,000 m^3 of water. The turbines were installed 200 m above sea level, and this also has a positive impact on their operational parameters. The short-term energy storage is connected with the long-term energy storage located in a valley by means of a pipeline. The response time for switching between the energy generation and storage functions is 30 s. The innovative nature of the project is determined by the short-term energy storage. The investment is fully automated. The hybrid power plant began its operation in 2018.

Keywords: innovative project, renewable energy, wind farm, pumped hydro storage, hybrid towers

1. Introduction

The operation of RES systems is characterised by significant random fluctuations of the amount of the produced electricity. On the one hand, this is the result of variable external conditions (geographical location and thus the resulting climatic conditions, impact of time, both per day and per year) and on the other hand, the changes in the demand for electricity among recipients. These fluctuations have a significant impact on the effectiveness of the systems. There are unpredictable, abrupt changes of power within the entire system, which may lead to its complete failure [1–3].

The problem cannot be ultimately resolved using the hybrid renewable energy source (RES) systems.

The integration of the selected storage technology with the source and the specific network is necessary. The proper selection is, above all, determined by such factors as storage capacity (kWh), output (kW), number of cycles (charge/discharge), service life (years/cycles), depth of discharge (%), response time (ms-min), efficiency (%), investment and operational costs [1–3].

Parameter	Power	Operating time	Life cycle	Energy density	Power density	Cycle efficiency	Response time
Unit	**[MW]**	**[h]**	**[years]**	**[Wh/l]**	**[W/l]**	**[%]**	**[s/min]**
Value	100–1000	4–12	30–60	0.2–2	0.1–0.2	70–85	seconds to minutes

Table 1.
Parameters of the pumped hydro storage technology.

At present, according to the report drawn up by the Electric Power Research Institute, the most frequently used large-scale electric energy storages in the world are pumped hydro storages. They constitute 95% of all large-scale storage facilities [2–7]. They are characterised by very good parameters, particularly in the case of storage of significant energy resources in the long-term scale.

However, it is necessary to bear in mind the limitations in their applicability, resulting from their location. If at least one of the reservoirs is a natural reservoir, it facilitates significantly the implementation of the enterprise and reduces its costs. The energy storage efficiency in the case of the pumped hydro storage ranges between 65 and 85%, but of particular significance is the fact that the response time of such a solution does not exceed several minutes and is often limited to seconds.

Table 1 includes information regarding the parameters of the pumped hydro storage (PHS) technology [2, 7].

In the case of storage of energy coming from renewable sources, it is advantageous to introduce the integrated storage systems, often on two levels, the short-term and long-term, which fulfil the role of buffer systems and thus ensure the reliability of power supply. The introduction of the short-term battery-operated energy storage is a good solution. At present the manufacturers of RES systems make attempts at integrating them with energy storages in such a manner as to secure the energy excess on a current basis [1, 2, 6]. An interesting case of such a solution is the Greek Icaria which relies on the wind-water system with two water storages (3.1 and 1 MW), located in different towns [8].

In the considered solution, the role of short-term energy storage is fulfilled by water reservoirs in hybrid wind turbine towers. They have many advantages, especially important in RES systems (Chapter 2.3): can be used for conventional and renewable sources, have long life cycle (50 years) with deep discharge, possible applications of fresh and salt water, lower investment costs due to standardised solutions and power plant concept (economic factor) and environment friendly (ecological factor). The system has been operating for only 7 months, so the time is not long enough for an exact evaluation.

2. The Gaildorf project

2.1 Environmental conditions

In the year 2011, the Gaildorf town initiated a discussion forum related to the natural power storage project. The project gained support of the local community and Bundesministerium für Umwelt, Naturschutz und Nukleare Sicherheit (BMUB), which supported the RES technological initiative in Gaildorf from funds coming from the ecological innovation programme in the amount of EUR 7,150,000 [9].

It was necessary to carry out multi-criteria research aimed at the confirmation of the selected renewable energy source solution both from the point of view of ecology and endangered species and detailed results of measurements regarding the wind energy availability [10–14].

While planning the location of wind farms, it is actually the evaluation of resources which seems to be the most difficult issue. It involves the evaluation of climatic conditions and the roughness of the terrain.

Another factor which remains highly significant with regard to the effectiveness of conversion is the impact of design parameters of the wind farm. However, the most important issue is the wind speed. The wind power as a function of its speed is expressed by the following relationship [10, 14, 15]:

$$P = \frac{1}{2}\rho A v^3 \tag{1}$$

where P is the wind power, A is the rotor blade area in m^2, v is the wind speed in m/s and ρ is the air density in kg/m^3.

With regard to the wind potential in the region in question, all the following European requirements are respected:

- The duration of the wind speed monitoring in the selected location should not be shorter than 1 year.

- The measurements should be conducted at a minimum height of 30–40 m, whereby the extrapolation of results to greater heights is permissible. The extrapolation is performed by means of the power law, Eqs. (2) and (3), or the logarithmic law [14]:

$$\frac{v}{v_0} = \left(\frac{H}{H_0}\right)^{\alpha} \tag{2}$$

$$\frac{P}{P_0} = \left(\frac{H}{H_0}\right)^{3\alpha} \tag{3}$$

where H is the turbine height, other markings as before, whereby indexed values 0 are measuring values for extrapolation purposes:

- The assessment of the wind potential must take into account the topography (roughness classes, surface friction coefficient α) [10, 13], for instance, α is the surface friction coefficient, −0.14 for low grass, 0.25 for low buildings and 0.40 for a built-up area.

- The measurements must be performed by means of two wind metres at two different heights (the averaging time was determined at the level between 10 and 60 min).

The results of the conducted research regarding climate, local wind conditions and wind parameters, as well as the applied research method, measuring devices, certificates for the calibration of anemometers and also the duration of monitoring, and the manner in which the results of measurements were converted into long-term data should be contained in the report. This is the most important document which is the basis for the economic assessment [16]. As well as the aforementioned information, the report should also include the final estimation of the annual energy output (AEO). During the economic analysis of the feasibility of the wind project, the value of AEO is of key significance. The factor which determines the annual energy output ratio is the wind speed.

For the Weibull distribution, AEO is described by the following relationship [14, 17, 18]:

$$AEO = \frac{1}{2}\rho A v^3 \eta t \tag{4}$$

where η is the efficiency (aerodynamic, mechanical, electrical) and t = 8.760 h/year, other markings as before.

The wind farm in Gaildorf is located at a site with average wind energy availability, and this is shown in the diagram presented in **Figure 1**. As can be seen, winds with speeds which exceed 60 km/h are, in principle, present in the periods between October and March. Their total duration is only about 60 h per year [19], but the winds with the speed reaching up to 38 km/h (10.6 m/s) occur within 25.5 days a year, and such a speed is optimal for the typical characteristics of the wind turbine [15].

As the wind rose shows, the winds from the West and South-West directions have the biggest potential for the location in question. Based on the meteorological data, it can be concluded that there is a significant variation regarding the availability of wind resources in time [19]; therefore, there is a necessity to store energy for the purpose of ensuring the power supply stability.

For example, **Table 2** includes the daily information regarding the wind speed in the region under consideration, on February 3 and April 3, 2019, at 3 h intervals, based on information given in [19].

In the further 7-day forecast, in February, the estimated wind speed values ranged between 6 and 13 km/h and in April between 10 and 18 km/h, respectively [19].

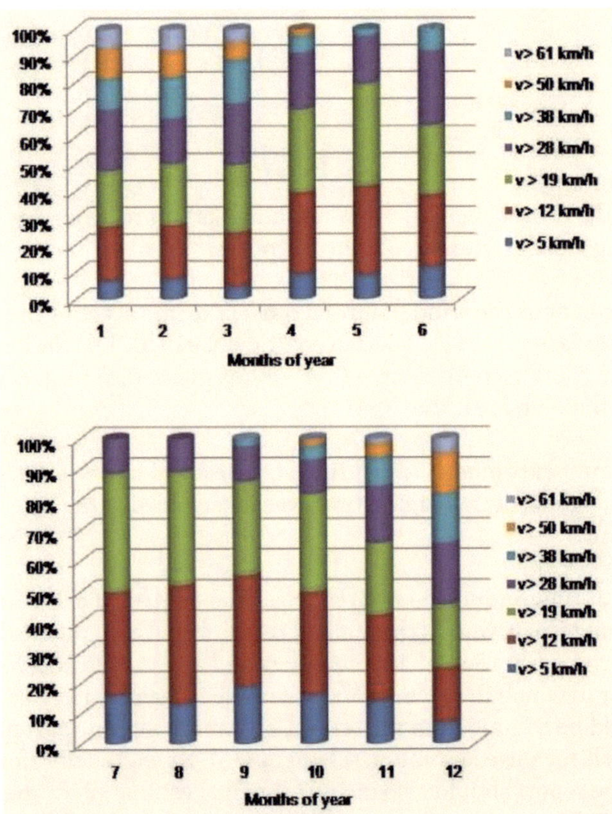

Figure 1.
The number of days with a specific wind speed during the respective months in Gaildorf, own study based on [19].

Day		Hour							
		2	5	8	11	14	17	20	23
February 3, 2019	Wind speed [km/h]	5–22	9–30	14–40	14–40	16–44	14–43	10–34	7–23
April 3, 2019		4–12	6–17	6–21	13–24	10–18	15–32	5–23	15–19

Table 2.
Wind speed in the region under consideration, on February 3 and April 3, 2019.

2.2 Wind turbines

In Gaildorf near Stuttgart, Bavaria, a non-standard wind-water energy solution was applied. A wind farm was established there in the year 2017 (**Figure 2**). The power of each of the turbines, with a rotor diameter of 137 m, amounts to 3.4 MW, which gives 13.6 MW in total for the entire system [20–23]. The farm consists of four electrically controlled turbines characterised by high performance. The hubs of GE 3.4-137 turbines are located in towers at a height ranging between 155 and 178 m. One of the turbines, since the moment of its installation, has been the highest turbine in the world. The total height of its tower and rotor wing reaches 246.5 m [20, 21, 23]. The heights of towers increased by 40 m are a consequence of their unique hybrid design (Chapter 2.3).

Additionally, the electricity production potential of the turbines was increased by their location on Limpurg Hills (the Swabian-Franconian Forest). In the case of the above-mentioned turbines (as marked in **Figure 3**), this allowed the following total heights to be obtained in relation to the sea level: W2, 501.5 m above sea level; W3, 489.5 m above sea level; W4, 489.5 m above sea level; and W5, 485.5 m above sea level [9, 22].

As has been mentioned in [10], an increase in the height at which the turbines operate is advantageous, as each additional metre of height of the location of the hub on the tower contributes to an increase in the annual energy output from 0.5 to 1%. The considerations must also take into account the impact of the ground roughness class [14].

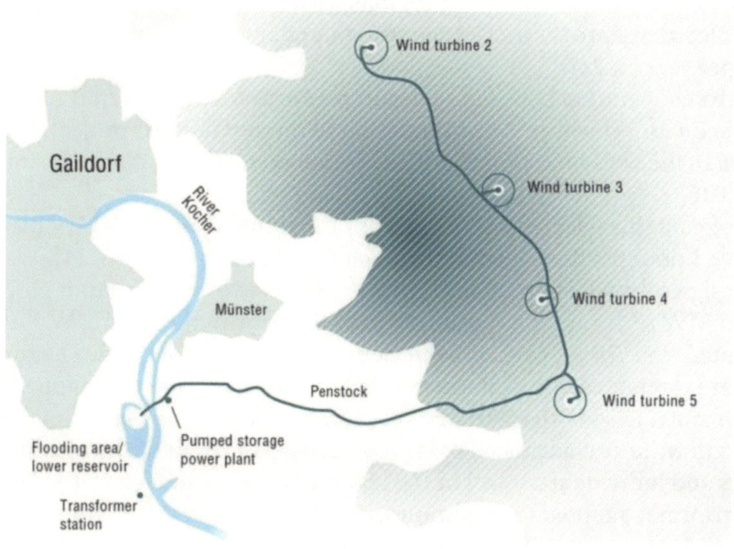

Figure 2.
Location of the wind-water investment in Gaildorf—the arrangement of wind turbines (Credit: Max Bögl).

Figure 3.
GE wind turbine from the 3 MW platform (Credit: GE Renewable Energy).

Parameter	Change in wind speed [%]	Change in wind power [%]
Change in height from 25 to 50 m	10	35
Change in height from 10 to 50 m	25	99

Table 3.
Change in wind speed and power with height.

Table 3 presents the impact of the turbine height parameters on the wind speed and its available power [14].

The significant heights at which the rotor hub is mounted also contribute to the reduction in turbulence and thus ensure a more stable operation of the turbine [10]. This is particularly important in the regions with quite small wind energy potential. The total electricity production by the system under consideration ensures about 42 GWh per year [9, 22].

With the cooperation between such companies as Max Bögl Wind AG and GE Renewable Energy Onshore Wind Deutschland on the Gaildorf project, the first wind farm in the world integrated with a hydroelectric power plant was initiated in the year 2016.

Table 4 contains selected specifications of turbines from the 3 MW platform, and **Figure 3** presents the GE wind turbine from that platform [16, 24].

The design solutions for turbines from the GE Platform with a unit power from the 3 MW series are the continuation of those from the GE 2 MW Platform, implemented since 2004; thus, proven elements have been applied. The turbines in question are three-bladed turbines mounted on a steel tubular tower, operating in the horizontal axis, with a variable speed, active control of the optimal deviation in relation to wind and advanced control of loads, as a result of the measurement of stresses and individual control of the blade angle of inclination [16]. The GE 3 MW Platform is adapted to operation with a broad spectrum of wind speeds, at standard and extremely low temperatures. It includes the 3.4 MW-137 m turbine which has best proven its value in class III. The direction of rotations of the rotor is clockwise. The solution involves the use of the electric adjustment of the drive and

Parameter	Value	Unit
Wind turbine power	3.2–3.8	MW
Rotor diameter	103, 130, 137	m
Annual electricity generation from wind	42	GWh
Turbine hub height above ground	150, 175, 178, 5, 199, 223	m HH

Table 4.
Design and operational parameters of GE wind turbines from the 3 MW platform.

Specifications of turbines of platforms II and III vs. wind class	Class no.	Wind speed	Wind classification
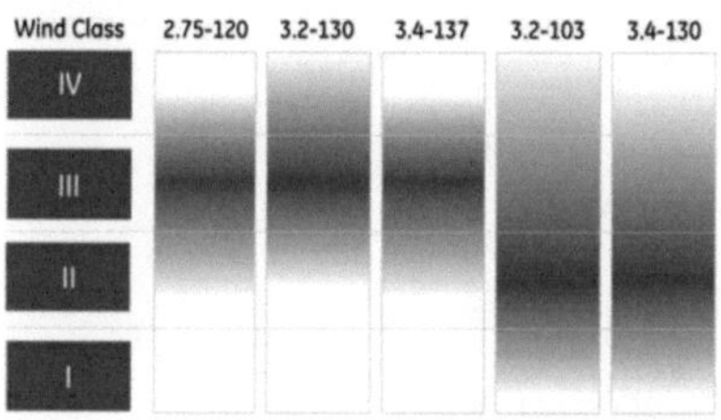	IV	≥29.0–32.0	Violent, hurricane wind
	III	≥25.0–28.0	Hurricane wind
	II	≥21.0–24.0	Windstorm
	I	≥17.0–20.0	Violent wind

Table 5.
Specifications of wind turbines, taking into account their applications in specific wind classes (Credit: GE Renewable Energy).

the aerodynamic brake. The internal and external anticorrosion protection has been taken into account. Under normal operating conditions, the sound power level does not exceed 106–107 dB (A) (the solution for noise reduction has been developed) [16].

Table 5 provides the specification of wind turbines, taking into account their applications in specific wind classes [16].

Owing to the modification of the gear box, control and aerodynamics, it was possible to increase the capacity, including the power factor, the annual energy production (AEP) and the scope of application (for 50 and 60 Hz frequencies). It must be emphasised that GE, including the wind turbines, also supplies the software dedicated to them. At present, in the Predix system of the company, the Digital Wind Farm software which collects information from various fields has become applicable. As a consequence of application of appropriate software (including the network connection and the communication interface), it is possible to obtain the optimal operation of each of the wind turbines.

The operation takes into account the GE condition monitoring system (CMS) and the Supervisory Control and Data Acquisition (SCADA) anomaly detection services, which complement the standard package. The advanced monitoring system (GE Global Research) allows for the discovery of any faults and anomalies of the system before they occur [16, 24].

As a result of modernisation work regarding the electric infrastructure, it was generally possible to limit capital expenditures (CAPEX) for the output power. It turned out in effect that in parallel to an increase in the energy potential, it was possible to reduce the outlays for production and operation. The costs of fuel were reduced, and the emission of CO_2 into the environment also became smaller [16].

Figure 4 shows the specification of the GE 3 MW platform [24]. The presented models are characterised by the same drive system and electric system whereby they are subject to modification on a current basis in order to ensure the highest energy

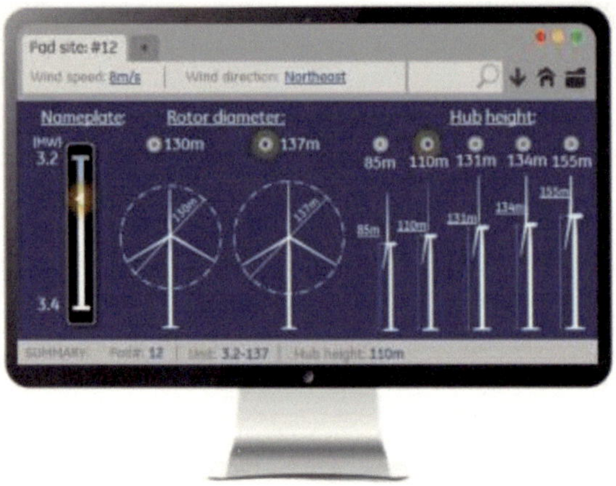

Figure 4.
The specifications of wind turbines from the GE 3 MW platform (Credit: GE Renewable Energy).

Figure 5.
Installation of wind turbine blades on the tower in Gaildorf (Credit: Max Bögl (Holger Hessenthaler)).

generation possible. The GE 3.4-130 wind turbine has an annual energy production (AEP) higher by up to 30% than the previous version, i.e. 3.2-103 [16, 24].

Figure 5 presents the fixture of the GE 3.4-137 wind turbine rotor blades (GE Renewable Energy Onshore Wind Deutschland) in Gaildorf. During the installation, the EF 300 Plus electric torque multiplier (Alkitronic) was used [25]. The device is characterised by a small size; therefore, it proves its value in the case of bolted joints at spatially confined locations. It ensures consistent and precise projection of a selected torque owing to the intelligent, processor-controlled electronics. High-quality screws are used in the discussed solutions. Owing to appropriate procedures, it was possible to tighten the screws in four Gaildorf wind turbines with a torque of 550 Nm plus 180° (1248 screws in total). This process was performed by means of an electric torque multiplier; thus, the joints could be made much faster than in the case of using hydraulic systems [25].

2.3 Hybrid towers: innovative energy storage

The innovative pilot system applied in Gaildorf, in connection with the introduction of the Max Bögl Wind AG project, uses the foundations of wind turbine towers as upper reservoirs (short-term energy storage) in the pumped hydro storage solution.

On the columns of towers, it is possible to separate appropriate sections (**Figure 6**).

1. The hybrid tower is based on the ring concrete foundation cast in situ. This ensures the transfer of significant loads coming both from tower's own weight and transferred to the ground by the wind force.

2. The highly efficient concrete elements are manufactured in series. It is necessary to ensure particular accuracy of workmanship of the elements as a result of precise grinding with computerised numerical control (CNC).

3. For the prestressing of the concrete tower, an external system for high-strength steel was used. It takes over the tension forces of the tower, which prevents the creep of concrete elements.

4. Adapter is a connecting element between two components of the tower: the one made of concrete and the one made of steel. The integrated support elements enable the placement of a steel pipe, and at the same time, the pipe adapter ensures prestressing forces for the concrete tower.

5. Steel section is a flexible construction which consists of several segments in order to facilitate the transport of the tower elements. The steel elements are manufactured at the factory owned by Max Bögl Wind AG, Sengenthal.

The foundations of the hybrid towers function at the same time as water reservoirs which ensure the energy capacity at the level of 70 MWh. The tower, which is built of modules, is provided with reservoirs along a section running from the ground to the height of 40 m: the vertical active reservoir with a diameter of 16.8 m and height of the head up to 31 m and the passive reservoir with a diameter of 63 m and depth ranging between 8 and 13 m. In total, it is possible to store 160,000 m^3 of water in four towers [9, 22].

The active reservoirs, which consist of 27 prestressed concrete rings, were built using the LTM 11200-9.1 (Hercules) mobile 9-axis crane owned by Max Bögl and provided with a ballast weighing 202 tonnes. After the rings (with a diameter of

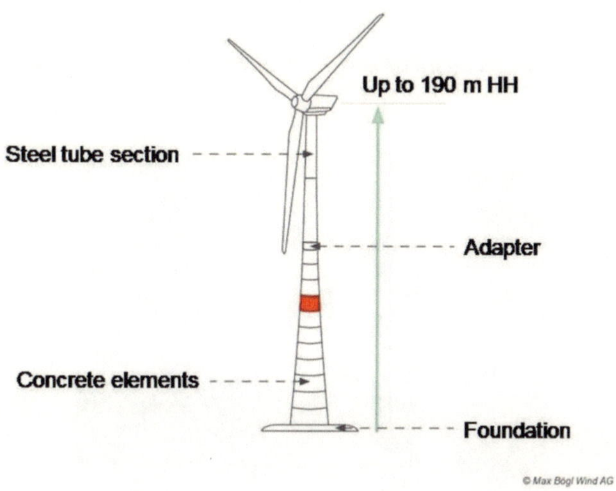

Figure 6.
Structural sections of the turbine tower (Credit: Max Bögl).

16 m and height of 1.5 m) were mounted on one another, they were secured with steel ropes [26].

Additionally, it was necessary to use the Liebherr 630 EC-H70 self-erecting slewing crane for installation works performed at the height of 190 m [26].

The components of the towers are ultimately put together at the construction site (**Figures 7–9**) [9, 26].

The lower part of the hybrid tower including the short-term storage is presented in **Figure 10** [9]. At the same time, this is the upper reservoir of the hydroelectric power plant.

Among the unquestionable advantages of upper basins located in the foundations of the towers, it is necessary to mention:

Figure 7.
Putting together concrete rings at the construction site by means of the Liebherr 630 EC-H70 self-erecting slewing crane (Credit: Liebherr Plant Ehingen GmbH).

Figure 8.
Hybrid tower on the concrete foundation (Credit: Liebherr Plant Ehingen GmbH).

Figure 9.
Installation of the internal steel tower by means of a crane (Credit: Max Bögl).

Figure 10.
Lower part of the tower including the short-term storage (Credit: Max Bögl).

- The reduction in the investment costs owing to standardised solutions
- The standardised concept of the power station
- The minimum interference in the landscape
- The possible applications with fresh and salt water
- The long service life (50 years) at deep discharge
- The possibility of combinations for applications other than RES

Figure 11 presents wind turbines in Gaildorf mounted on towers [22].

Ultimately, the wind turbines were integrated with the network in the Gaildorf project in spring (2018). The wind farm began its operations [9].

The presented energy storage concept in the short-term scale is a completely new solution. The water to reservoirs in towers was supplied by means of the DN 1600–2500 diameter pipeline.

Figure 11.
Wind turbines on towers in Gaildorf (Credit: Max Bögl).

A particularly important factor for the operation of the system is the time of switching between the energy generation function and its storage function, which is 30 s [21]; this ensures quick response of the system to the current needs. The ability to store energy is characterised by 80% capacity.

The integration of wind turbines and the hydroelectric power plant as the energy storage is aimed at gaining independence from environmental and climatic factors.

2.4 Hydroelectric power plant

The upper water reservoirs (in the tower foundations) are combined with the lower one located in the Kochertal valley at a distance of 3.2 km (200 m in the vertical line) (Gaildorf-Unterrot) by means of a throttle. The design by Max Bögl Wind AG in its current form allows for differences in heights ranging between 150 and 350 m between the lower reservoir and the bases of wind turbine reservoirs [22].

For the needs of the investment, special DN 1800/DN and 1600/DN polyethylene discharge pipes (manufactured by Egeplast), running from the reservoirs, were designed. They were laid using a platform-based machine—PiPECrawler—intended for the installation of thermoplastic pipelines on an industrial scale (**Figure 12**). The machine guarantees three times higher operating speeds than those proposed in standard solutions and allows any unevenness and obstacles in the topography to be taken into account much more effectively.

Also its impact on the environment is reduced to the minimum.

The innovative PiPECrawler solution, which is an internal patent developed by Naturspeicher GmbH of the Max Bögl corporate group, has been honoured with the first "Development for Industry" award in the year 2019 [27].

Figure 12.
The operation of the PiPECrawler in Gaildorf (Credit: Max Bögl).

The Egeplast polyethylene pipes had been installed since 2018. Then a steel distribution pipeline "trifurcator" was installed. This is a branch pipe which allows for the reaching of three turbine sets in the hydroelectric power plant. They were installed in the Gaildorf hydroelectric power plant in autumn (2018). This scope of works was the responsibility of Bilfinger VAM Anlagentechnik GmbH from Upper Austria [28, 29].

The bottom reservoir of the hydroelectric power plant is characterised by the natural appearance, owing to which it does not affect the landscape visually. In its case, the issue of water management must be considered comprehensively. It is not completely filled with water. It is necessary to maintain the spatial reserve of 30,000 m^3 on a constant basis, which ensures safety in the event of a flood, as the lower reservoir also fulfils the role of a retention reservoir.

The works related to the construction of the hydroelectric power plant were completed at the turn of 2018 and 2019. At the investment stage, it is optionally possible to include the storage capacity of 16, 24 and 32 MW in the system [22].

It must be emphasised that the hydroelectric power plant is fully automated.

In the case of demand for electricity, water flows downwards from the basins in the turbine towers, supplying the hydroelectric power plant located in the valley. During the peak demand for electricity, the water from the upper reservoirs is released to the lower Gaildorf-Unterrot reservoir by opening a special valve. Its flow drives the water turbines. When there is electricity surplus, the water from the lower reservoir is pumped through pipelines to the upper reservoirs in the foundations of the towers on the hill. These reservoirs are to function as a huge storage which complements both the networks and the wind turbines. The flow speed in the Gaildorf hydrosystem is estimated at 9.5 m^3/s. Electricity will be generated as a result of operation of wind turbines and water turbines.

During analysis of operation of the hydroelectric power plant, the evapotranspiration of water, which covers the process of volatilisation of the water into the atmosphere through plant transpiration and sublimation, was taken into account [30, 31].

Three reversible Francis turbines with the total power of 16 MW were applied in the Gaildorf solution. They are mainly used to drive the power generators in hydroelectric power plants with an appropriately high water fall. The special matching of the Francis turbines to the installation facility allows the efficiency exceeding even 90% to be guaranteed. The cost of manufacture and installation of the turbines is high, but their performance is very reliable. The Francis turbines are reversible devices. They can be applied in pumped hydro storages where the surplus of electric power is used to pump water from the lower reservoir to the upper one. The water

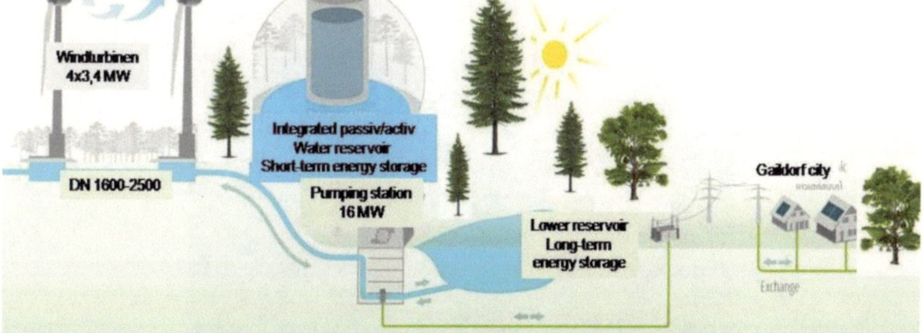

Figure 13.
Sketch of the wind-hydro system in Gaildorf (own study based on [22]).

turbines in this project are within the scope of responsibility of a German company bearing the name Voith Hydro [32, 33].

South of the hydroelectric power plant, a substation will be built to provide the connection with the existing 110 kV line of the BW operator.

The draft of the Gaildorf system, which covers all the components, is provided in **Figure 13**.

3. Conclusions

1. The faster and faster development of electricity generation technologies based on renewable sources and their popularisation requires system solutions which enable energy storage both on a short-term and long-term basis. Selected energy storage technologies should be integrated both with the RES source and the specific network.

2. In the case of the hybrid system, synergies between the technologies are of particular importance. The costs of planning and construction of the whole complex of facilities are lower than in the case of separate consideration of operation of the respective components, which surely is a fact that cannot be ignored.

3. The presented wind-water solution—the Gaildorf pilot scheme—is innovative in its nature and has not been used before. The water batteries designed and made by the Max Bögl corporate group located in the wind turbine towers are a kind of a short-term storage. It guarantees the highest efficiency of conversion of wind energy into electricity. It is also an element which consolidates the wind turbine with the pumped hydro storage in the valley (**Figure 14**) [22].

4. The introduction of a standardised solution, i.e. reservoirs in the foundations of turbine towers, is characterised by lower dependence on the location and higher reliability. The water battery is a new solution, a flexible miniaturised pumped hydro storage, which is an intermediate stage between the renewable energy source and storages on a broad scale. The standardisation of short-term storages does not require a single large reservoir; therefore, it does not require the adaptation of the facility to the location and the adaptation of the investment to the reservoir parameters. This often causes the reduction in the storage potential as was the case with the hybrid investment on the island of El Hierro. Neither does it require the use of a standard battery storage to complement the accumulation potential of the system.

Figure 14.
Arrangement of the components of the wind-hydro pilot project in Gaildorf (Credit: Max Bögl).

5. It must be emphasised that the innovative energy solution introduced in Gaildorf required the involvement of teams of experts from Germany and abroad in the works. The energy concepts developed by them must be considered both in the aspect of the energy transformation and the global climate change.

The broadly understood cooperation on the Gaildorf project and the final integration of the effects obtained by the teams will most surely contribute to the achievement of the objective set by Germany, i.e. the production of 45% of electricity from RES already in the year 2027.

6. The Gaildorf project combines energy storage with its generation by RES and control systems. The presented investment is a confirmation of the thesis by Jérôme Pécresse (senior deputy president of GE & CEO, GE Renewable Energy) about the growing interest in developments regarding RES, complemented by energy storages. Only such solutions guarantee rapid responses of the system to load changes and also full reliability.

Another project involving the integration of renewable energy sources, solar and wind energy and energy storages, is developed in the Indian Kadapa Hybrid Park, which will significantly improve the network stability.

Conflict of interest

The author declares no conflict of interest.

Notes/thanks/other declarations

Special thanks to Nancy Fürst from Max Bögl, Agathe Lefevre De la Houpliere from GE Renewable Energy and Meryem Yaman from Liebherr-International Deutschland GmbH for their photo contribution.

Author details

Grażyna Frydrychowicz-Jastrzębska
Department of Electrical Engineering, Poznan University of Technology, Poznań, Poland

*Address all correspondence to: grazynajastrzebska@op.pl

References

[1] Akhil AA, Huff G, Currier AB, Kaun BC, Rastler DM, Chen SB, et al. Electricity storage handbook in collaboration with NRECA. Sandia Report No. SAND 2013-July 5131; United States; 2015. DOI: 10.2172/1170618

[2] Decourt B. Electricity Storage, Leading the Energy Transition Factbook. Gravenhage: SBC Energy Institute; 2013

[3] Infield D, Freris L. Renewable Energy in Power Systems. 2nd ed. Chichester: Wiley; 2018. 350p. ISBN-13:978-1118649930

[4] Foley AM, Leahy PG, Li K, McKeogh EJ, Morrison AP. A long-term analysis of pumped hydro storage to firm wind power. Applied Energy. 2015;**137**:638-648. DOI: 10.1016/j.apenergy.2014.07.020

[5] GE Renewable Energy. Implementing Sustainable Business Models for Hydro Storage. Paris: GE Renewable Energy; 2019

[6] Frydrychowicz-Jastrzębska G. El Hierro renewable energy hybrid system: A tough compromise. Energies. 2018;**11**(10):2812. DOI: 10.3390/en11102812

[7] Locatelli G, Palerma E, Mancini M. Assessing the economics of large energy storage plants with an optimisation methodology. Energy. 2015;**83**:15-28. DOI: 10.1016/j.energy.2015.01.050

[8] Tselepis S. Storage for non-mainland grids and other special situations. The future role of energy storage in South Eastern Europe. In: Proceedings of the Enlargement and Integration Action Workshop; 21-22 October 2014; Tirana Albania. Luxembourg: Publications Office of the European Union; 2014. pp. 7-9

[9] Info Max Bögl Gaildorf News [Internet]. 2019. Available from: https://www.mbrenewables.com/en/pilot-project/ [Accessed: 2019]

[10] Frydrychowicz-Jastrzębska G, Bugała A, Janczak D, Dach J, Zaborowicz M. The efficiency of the modern wind turbine as a function of the wind energy potential resulting from the construction. In: Proceedings of the 3rd International Conference on Energy and Environment: Bringing Together Engineering and Economics (ICEE'17), 20-30 June 2017; Porto, Portugal; 2017. pp. 335-342. ISBN: 978-972-95396-9-5

[11] Wizelius T. Wind Power Projects: Theory and Practice. 1st ed. London/New York: Routledge; 2015. 230p. ISBN: 9781138780453

[12] Wizelius T. Design and implementation of a wind power project. In: Letcher T, editor. Comprehensive Renewable Energy. United Stated of America: Elsevier; 2012. pp. 391-430. DOI: 10.1016/B978-0-08-087872-0.00215-8. ISBN: 9780080878720

[13] Gipe P. Wind Energy Basics: A Guide to Home and Community-Scale Wind-Energy Systems. 2nd ed. Vermont: Chelsea Green Publishing; 2009. 192p

[14] Gipe P. Fundamentals of Wind Energy (Presentation) [Internet]. Available from: wind-works.org [Accessed: March 20, 2019]

[15] Jastrzębska G. Energia ze Źródeł Odnawialnych i jej Wykorzystanie. 1st ed. Warszawa: Wydawnictwa Komunikacji i Łączności; 2017. 334p. ISBN: 978-83-206-1983-6

[16] GE Renewable Energy. GE's 3 MW Platform Powerful and Efficient GE

Renewable Energy Trademark of General Electric Company Copyright © 2017 General Electric Company [Internet]. 2017. Available from: https://www.ge.com/content/dam/gepowerrenewables/global/en_US/downloads/brochures/wind-onshore-3mw-wind-turbine-platform-gea32208b-r1.pdf [Accessed: April 18, 2019]

[17] Borja MA, Lagunas J. Miscalculations on the estimation of annual energy output (AEO) of wind farm project. Energy Procedia. 2014:698-705. DOI: 10.1016/j.egypro.2014.10.225

[18] St. Martin CM, Lundquist JK, Clifton A, Poulos GS, Schreck SJ. Wind turbine power production and annual energy production depend on atmospheric stability and turbulence. Wind Energy Science. 2016;**1**(2): 221-236. DOI: 10.5194/wes-1-221-2016

[19] Weather [Internet]. 2006-2019. Available from: https://www.meteoblue.com/en/weather/ [Accessed: 2019]

[20] Colthorpe A. World's Tallest Wind Turbine Gets 70 MWh of Pumped Storage Near Stuttgart. Energy Storage [Internet]. 2017. Available from: https://www.energy-storage.news/news/worlds-tallest-wind-turbine-gets-70mwh-of-pumped-storage-near-stuttgart [Accessed: April 18, 2019]

[21] Dvorak P. Max Bögl Wind Puts Turbine on the Tallest Tower, 178 m. Blade Tip to 246.5 m Windpower Engineering & Development [Internet]. 2017. Available from: https://www.windpowerengineering.com/mechanical/towers-construction/max-bogl-wind-puts-turbine-tallest-tower-178m-blade-tip-246-5m/ [Accessed: April 18, 2019]

[22] Info Gaildorf News [Internet]. 2019. Available from: https://www.max-boegl.de/en/ [Accessed: 2019]

[23] Casey T. GE Forges Ahead With Worlds First Wind+Hydro+Storage Project. Clean & Technica [Internet]. 2016. Available from: https://cleantechnica.com/2016/11/14/ge-forges-ahead-worlds-first-wind-hydro-storage-project/ [Accessed: 2019]

[24] 3 MW Wind Turbine Platform [Internet]. 2019. Available from: https://www.ge.com/renewableenergy/wind-energy/onshore-wind/turbines/3mw-platform [Accessed: 2019]

[25] World Champion of Height in Wind Assembly. Case Study Max Bögl, Gaildorf, Germany [Internet]. Available from: https://www.alkitronic.com/fileadmin/user_upload/referenzen/Projekte/Downloads_Case_Studies/englisch/alkitronic_Casestudy_Naturstromspeicher_EN.pdf [Accessed: April 16, 2019]

[26] Höchste Windkraftanlagen der Welt: Max Bögl baut mit Liebherr-Mobilkran Modernen Energiespeicher [Internet]. 2017. Available from: https://www.liebherr.com/de/deu/aktuelles/news.pressemitteilungen/detail/h%C3%B6chste-windkraftanlagen-der-welt-max-b%C3%B6gl-baut-mit-liebherr-mobilkran-modernen-energiespeicher.html [Accessed: April 18, 2019]

[27] Method Used to Lay Pipe for Naturstromspeicher Gaildorf Energy Storage Pilot Project Wins Award. Hydro Industry FAQS and Resources [Internet]. 2019. Available from: https://www.hydroworld.com/articles/2019/02/method-used-to-lay-pipe-for-naturstromspeicher-gaildorf-energy-storage-pilot-project-wins-award.html [Accessed: March 18, 2019]

[28] Bilfinger liefert Verteilrohrleitung für Wasserbatterie-Projekt [Internet]. 2018. Available from: https://www.bilfinger.com/media/news/bilfinger-liefert-verteilrohrleitung

fuer-wasserbatterie-projekt/ [Accessed: March 18, 2019]

[29] Der Naturstromspeicher GmbH Brochure [Internet]. 2018. Available from: https://naturspeicher.de [Accessed: March 18, 2019]

[30] Bakken TH, Killingtveit Å, Alfredsen K. The Water Footprint of Hydropower Production—State of the Art and Methodological Challenges. 1st ed. Weinheim: Wiley; 2017. DOI: 10.1002/gch2.201600018

[31] Hoekstra AY, Chapagain AK, Aldaya MM, Mekonnen MM. The Water Footprint Assessment Manual. Setting the Global Standard. 1st ed. London, Washington: Earthscan Ltd.; 2011. 203p. ISBN: 978-1-84971-279-8

[32] Innovative Combination of Wind and Water–Voith to Deliver Pump Turbines for First Combination of Wind and Pumped Storage Power Plant [Internet]. 2016. Available from: http://voith.com/corp-en/news-room/press-releases_72434.html [Accessed: March 18, 2019]

[33] Reidy R. Voith to Deliver Pump Turbines for Wind/Pumped Storage Power Plant. World Pumps [Internet]. 2016. Available from: https://www.worldpumps.com/power-generation/news/voith-to-deliver-pump-turbines-for-windpumped/ [Accessed: March 18, 2019]

Chapter 9

Hybrid Energy-Based Chilling System for Food Preservation in Remote Areas

Edwin Mohan, Saranya Nair Mohan and Joseph Sekhar Santhappan

Abstract

The milk processing and preservation is a fast growing business in developing countries and it is facing problems due to high energy cost and environmental concerns in using conventional energy sources. Since the selection of feedstock and conversion technologies, appropriate research must implement renewable energy-based technologies to promote a constant flow of energy services. In this chapter, the focus is on implementing cooling technologies, using locally available energy sources such as biomass, biogas, gobar gas, which is going to be popular in the near future. The renewable energy sources can be used alone or in combination to run the generator of the vapor absorption system. Sufficient study is not available for hybrid energy systems, with the combination of locally available energy sources, focused in this study. Therefore a systematic analysis is needed to find the appropriate mixing of various renewable energy sources to meet the cooling requirements in any region to implement the complete renewable energy-based cooling system. The effect of variations in the combination of renewable energy sources on the overall system COP has been studied. Based on the maximum system performance and best economic performance, suitable combinations that can be preferred in various regions are predicted.

Keywords: hybrid energy, renewable energy, bio-energy, cooling system, remote area

1. Introduction

In developing countries, the gap in the quality life of the urban and rural areas is severely affected by the lack of various facilities which require electricity. Conventional energy sources such as electricity and diesel is not available in many rural regions and the grid connection and transportation of the fossil fuels are very difficult in rural India, because of the large capital investment required for the electrical infrastructure. As a result, marginal and small farmers face difficulty preserving the milk, fruits and vegetables in remote places. However, poor infrastructure, including the lack of integrated cooling facilities (cold chain), has retarded the growth of the milk and food processing industry [1, 2]. Due to a lack of proper storage and transit facilities, about 22% of agricultural produce, especially fruits and vegetables, are spoiled, and the loss is estimated to be worth about Rs. 330 billion. This spoilage could be prevented at the local village level by providing cooling units for short term preservation. Most of the cooling units are operated by vapor compression refrigeration systems. These systems need electrical power, which is not readily available in remote areas. Moreover, the cooling facility has to be located in close proximity of the source of the raw materials, especially perishable agricultural products, because it would help in reducing post-harvest losses and wastes. Since India is blessed with perennial solar energy and surplus bioenergy resources. Implementing cooling technologies, using locally available energy sources such as biomass, biogas, gobar gas, is going to be popular in the near future. With the abundant renewable energy sources in remote areas, it is indeed feasible for biomass, biogas, gobar gas and solar energy to be suitably integrated to operate a thermal driven absorption cooling system. With this the energy deficiency for food preservation could be solved. This unique factual situation motivated the researcher to carry out this work.

When considering a renewable energy system for rural requirements, it is important to design systems that are reliable and require little maintenance, as frequent and guide repairs and replacements might not be feasible. Using a stand-alone renewable energy system, such as solar, biomass and biogas conversion system, it is not possible to produce energy in the off-peak period and transportation of supply of bio-energy sources in a rural area is not easily possible. So integration of locally available energy sources in rural areas can be used to overcome these drawbacks. A hybrid energy system generally consists of a primary renewable source working in parallel with a standby secondary non-renewable module. Instead of using single stand-alone units, larger hybrid systems would be more cost-competitive for remote communities [3, 4]. Hybrid energy system includes several (two or more) energy sources, with appropriate energy conversion technology, connected together to feed power to local load and it would increase the energy reliability and overall efficiency.

In this research, modeling and simulation of hybrid energy (biomass/biogas/gobar gas/solar)-based cooling system has been successfully made using MATLAB software. An improvement in overall system COP (COP_{OS}) and best economic viability of the new renewable energy combination has also been demonstrated in this work.

A number of studies associated to the viability of cooling systems with hybrid energy and renewable energy have been carried out with respect to its application, possibilities, enactment, optimization, combination with another kind of energy conversion techniques and so on. Hybrid energy system generally consists of a primary renewable energy sources working in parallel with a standby secondary non-renewable module. This hybridization reduces 50% of the excess energy and also cost-competitive for remote communities [5, 6]. In hybrid system, two or more power generation sources are balanced to each other's strength and weakness [7, 8]. Suitable combinations of solar, biogas and biomass energy sources to meet the energy demands in order to improve the environment and socio-economic conditions in remote places [9]. Moreover the hybrid combination of renewable energy systems can be a cost-effective, alternative to grid extension, sustainable, techno-economically viable and environmentally sound system [10]. Cost, efficiency, social acceptance, reliability, and potential demand were the important factors to design the hybrid energy model [11, 12]. Even though several kinds of energy resources are obtainable in the villages of developing nations, a sole energy resource cannot meet the energy constraint [13]. Since, inadequate bioenergy resources, solar energy system is nominated to compensate for energy requirement.

Solar and biogas driven aqua-ammonia refrigeration system can be used for improving the quality of milk preservation at remote dairy farms, it shows an economically attractive [14, 15]. The energy demand can be matched with the supply of biogas by an appropriate choice of solar collector [16]. A hybrid cooling system with solar and biomass energy sources used to increase the overall system performance with 0.11. It shows that the performance has more consistency compared with conventional cooling system in remote places [17–20]. Nixon et al. [21] have reported that the hybrid energy system combined with solar and biomass energy sources which decreases the demand of biomass up to 25–30%, it results the constant energy security and reduce the requirement of land. Since, it has environment friendly with long payback period. The drying time of the solar-biomass drier has been reduced to 54–60% [22]. Wind energy conversion system combined with solar, fuel cell, hydro and biomass energy conversion system fulfill the remote consumer's electrification needs and guarantee a long-term energy autonomy, and minimize the consumption of imported oil and the corresponding environmental impacts. Moreover, this hybrid energy system is a better alternative for a wind-diesel system. MATLAB, response surface meta-models (RSM) and genetic algorithm (GA)-based analysis techniques are most widely used methods for optimize such type of hybrid energy system [23–28].

Standalone conventional electric power generation system retrofitted with renewable energy sources would be a viable option for power production at remote locations. It also stated that the PV/wind/hydro/diesel hybrid system shows the energy cost about $0.207/kWh [29]. Biomass generator is chosen as a major source of power in rural areas due to its high efficiency and cost effectiveness. The lowest energy cost observed is 1.85 Rs/kWh [30]. The hybridization of solar, wind, biogas, biomass, and hydro energy sources is to satisfy the viable electrification option in remote areas. It shows better economical solution and also a reliable option to moderate the present power crisis. It can have a good impression on enlightening the socio-economic situations of rural people [31]. Mixed integer linear programming model defines the optimum configuration of hydro/biogas/biomass/solar operated hybrid energy system with estimation of the economic diffusion levels of photovoltaic array area, and cost optimization [32–35].

Past literature shows that the most of the studies on the hybrid energy systems concentrated on the electrical energy production. More attention is needed for the effective operation of cooling and thermal power energy. Pecuniary environment of proposed combinations of available sustainable power resource frameworks are a significant issue. An epic blend of a few sustainable power source assets with further energy efficient feature is required to diminish the expense of the hybrid energy system. This can doggedness the issues to be challenged by the chilling amenity area in the inaccessible places and furthermore the 100% sustainable power source-based framework gives the ecological advantages.

1.1 Scope and aims of the work

Conventional energy-based vapor compression chilling system and stand-alone renewable energy-based electric power production are well-known technologies that are broadly used in developing countries. In disparity, hybrid thermal energy powered chilling system is a relatively novel notion. The conversion technology of bio-energy resources and solar power has been serious factor for the effective enactment of hybrid energy powered absorption chilling system to preserve the food and other products in remote areas, where the deprived electrical and transport facility. Thus, the primary study contained within this work contributes toward the selection of available renewable energy resources and requirement of cooling load, and mathematical design of a biomass, biogas and gobar gas energy conversion systems for the selected region. In view of this the ensuing objectives were delineated.

1. For different regions and locations, climatic conditions, including the availability of energy resources, requirement of cooling load, and so forth, are always changing. In order to efficiently utilize that the available renewable energy sources in these regions, Proper data collection and data fusing is needed for system process and thermal load requirement.

2. Detailed modeling and simulation analysis on biomass, biogas, gobar gas, solar energy conversion systems and suitable system with absorption chillers accordance with these energy resources.

3. Overall performance of the hybrid energy-based absorption cooling system with various energy proportions suitable to the study region-based on the nature of activities.

4. Economic studies of milk chilling system worked with existing available renewable energy resources.

5. Based on the techno-economic investigations, identification of the appropriate mixture of existing renewable energy sources for milk preservation, with higher overall performance of the system and minimum cost.

2. Materials and methods

2.1 Load/demand assessment

Agriculture-based industry has a vast significance in developing countries since the vital relations and interactions that it endorses between the industry and agriculture. Milk and food preservation covers a variety of yields from sub sectors including agriculture cultivation, farmstead animal farming and fisheries. Agro and milk-based food sectors are anticipated to show a key role in rural farming.

Data pertaining to the bio resources, no's of livestock and requirement of cooling load were composed from the local peoples, Govt. and non-Govt. agencies, etc., and the consistency of the data was inveterate from response received from local societies. Field studies based on household and direct discussion were carried out in the region to gather the information on obtainability of renewable energy resources, existing energy consumption, etc. The significant particulars composed as of the investigation are shown in **Table 1**. The main biomass energy resources (BM) are tapioca stalk, coconut shell, coconut coir pith, coconut rachis, wood pellets, rubber

Particulars	Survey details	Particulars	Survey details
No of villages	6	Geographical area of the region	26 sq km
Total population (Nos)	1750	No of households (Nos)	438
Density of livestock population (Nos)	160	Quantity of dung production (kg/day)	740
Biomass sources (kg/day)	390–640	Biogas sources (kg/day)	640–930
Gobar gas sources (kg/day)	660–880	Quantity of milk production (L/day)	1700

Table 1.
Details of survey data of selected regions.

seed, etc. Likewise the biogas (BG) and gobar gas (GG) resources are considered from the municipal solid waste and as of the populace of cows. The cooling requirement to keep the quality of milk produce also the requirement of heat load to work the cooling framework is to be resolved dependent on the yield of the milk, no of cows and the accessibility of the sustainable power sources.

The capacity of the production of milk in the isolated areas is calculated by the ensuing equation:

$$\begin{aligned} V_m = &(N_{cows} \times \text{milk produced/cow}) + (N_{buffalo} \times \text{milk produced/buffalo}) \\ &+ (N_{goats} \times \text{milk produced/goat}) \end{aligned} \tag{1}$$

where, V_m is the volume of the milk produced per day in L/day. Normally each cow produces 8–16 L of milk per day [36, 37]. N_{cows} is the number of cows/cattle.

The available mass of milk is determined by

$$m_m = V_m \times \rho_m \tag{2}$$

where, m_m is the mass of milk produced per day in kg/day, ρ_m is the density of milk in kg/m^3 (density of milk is taken as 1027 kg/m^3) [38].

2.2 System description

The schematic diagram of a proposed hybrid energy system is shown in **Figure 1**. The proposed hybrid energy-based cooling system consists of two major sub-systems. (1) Energy conversion system; (2) vapor absorption cooling system.

Energy conversion system converts the available energy from the feedstock sources to thermal energy required for supply of heat to the steam generator of a vapor absorption cooling system. The number of conversion devices like biomass gasifier, biogas plant, gobar gas plant and solar collector has been used for the energy conversion process. These devices are selected based on their availability of energy resources and the required cooling load, whichever the Lithium Bromide-water or water-ammonia refrigeration system has been selected as the cooling device. The entire cooling load for the milk chilling is from the evaporator of the vapor absorption cooling system.

The preliminary analysis states that the renewable energy resources such as biomass (BM), biogas (BG) and gobar gas (GG) can be taken as main energy resources. When the BM, BG, and GG are not adequate to meet the cooling requirement, solar energy (SO) has to be used beside with the aforementioned renewable energy resources. In most of the areas in the study region, there is a major deviation in solar intensity during 7–9 months in a year because of cloudy condition. Moreover the preliminary cost of the solar conversion system is also very high compared with other renewable energy conversion devices. Hence BM, BG and GG energy resources plays a major renewable energy resources and solar energy has been used where the aforementioned energy resources are not adequate to meet the

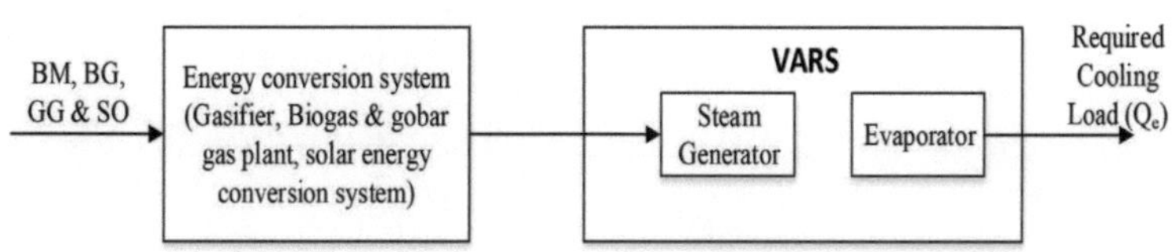

Figure 1.
Proposed hybrid energy-based cooling system.

energy constraint. Wood pellets, tapioca stalk, coconut stalk, coir nub, paddy straw, rice shell, coconut shell, etc., are used as energy harvests in biomass gasifier. Municipal solid waste and food waste from rural households are the resources of energy in the biogas plant. Cow dung is used as an energy yield in the gobar gas plant. A LiBr-water vapor absorption system considered as COP of 0.5–0.75, evaporator temperature of 4–9°C and generator temperature of 95–112°C [39].

2.3 Theoretical modeling

The strategies followed to analyze the hybrid system are depicted in a flow chart shown in **Figure 2**. The energy conversion has been determined, in light of the conversion efficacies of biomass (η_{cBM}), biogas (η_{cBG}) and gobar gas (η_{cGG}), and their qualities are taken as 0.46, 0.26, and 0.35 respectively [40].

The quality amount of agro squanders have been evaluated based on the accessible information. The heat load created as of the biomass gasifier (Q_{BM}) can be resolved from [18–20].

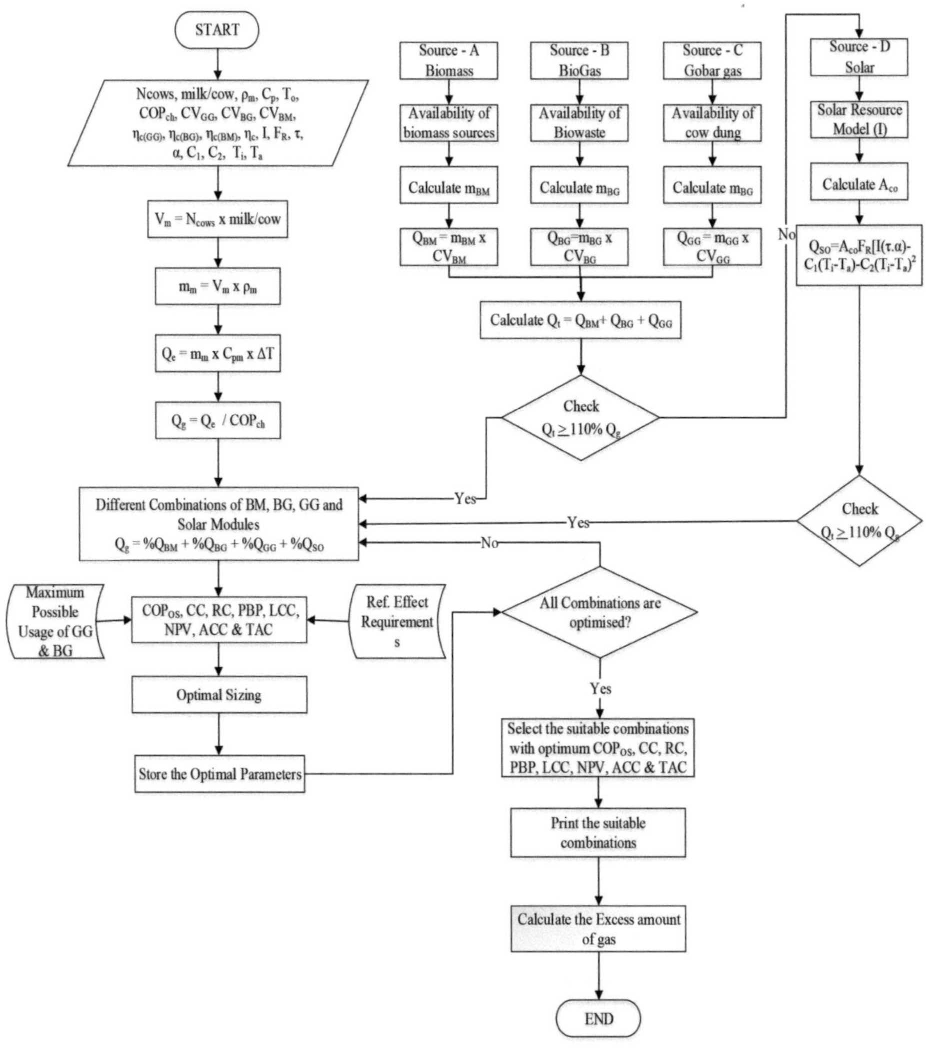

Figure 2.
Flow chart for the simulation of proposed hybrid energy system.

$$Q_{BM} = m_{BM} \times CV_{BM} \times \eta_{cBM} \tag{3}$$

The quantity of solid waste generated per family is taken as 3–6 kg per day [14, 15], and the amount of waste required to produce 1 m^3 gas is assumed as 23 kg [14, 15]. The heat produced from the biogas plant (Q_{BG}) is given as [18–20].

$$Q_{BG} = m_{BG} \times CV_{BG} \times \eta_{cBG} \tag{4}$$

Similarly, the biogas produced from cow dung (gobar gas) has been calculated from the cattle population and dung yield per animal. The biogas production from the cow dung has been worked out, based on the assumption that 4–7 kg dung per cow per day is produced [14, 15], and the amount of cow dung required to produce 1 m^3 gas is estimated as 12 kg [15]. The heat produced from the gobar gas plant (Q_{GG}) can be obtained from the equation [18–20].

$$Q_{GG} = m_{GG} \times CV_{GG} \times \eta_{cGG} \tag{5}$$

On a clear day, the average solar radiation is about 700 W/m^2, and the average solar conversion efficiency could be 50%. The useful energy output of a collector (Q_{SO}) in the steady state condition is calculated as follows [21]:

$$Q_{so} = A_{co}F_R\left[(\tau\alpha)I - C_1(T_i - T_a) - C_2(T_i - T_a)^2\right] \tag{6}$$

The entire system design is based on the cooling load of the selected region. The evaporator cooling load (Q_e) of the vapor absorption refrigeration system in the steady state condition, is obtained from the basic relation

$$Q_e = m_m \times C_{p,m} \times \Delta T \tag{7}$$

where, Q_e is the amount of cooling effect, delivered from the evaporator in kJ/kg, $C_{p,m}$ is the specific heat of milk in kJ/kgK, ΔT is the change in temperature of the milk in the evaporator.

The required amount of BM, BG and GG sources are considered based on the requirement cooling capacity and the heat load requirement to the steam generator in the vapor absorption refrigeration system. The required amount of refrigeration capacity is resolved dependent on the Eq. (7). The required heat load is determined dependent on the refrigeration capacity and the performance of the refrigerating system. The amount of heat load provided to the generator is to be calculated from the forthcoming relation [18–20].

$$Q_g = Q_e/COP \tag{8}$$

Where, Q_g is the heat supplied to the generator in kW. Assume COP of the absorption chiller is 0.5 [18, 19].

The overall system performance (COP_{OS}) of the proposed system is determined from the equation

$$COP_{os} = Q_e/(k.IA_{co} + x.m_{BM}CV_{BM} + y.m_{BG}CV_{BG} + z.m_{GG}CV_{GG}) \tag{9}$$

Where x, y, z, and k denote the proportion of biomass, biogas, gobar gas and solar energy in hybrid energy system.

The base energy source is considered as any of the single energy resource, and the mixture element is the combination of another two or three energy resources.

The mixture component ratio is changed from 25, 50, and 75% respectively. In all magnitudes the proportion of the energy resources in the blend component is kept steady for a specific blend.

MATLAB programming has been utilized to simulate the hybrid energy-based chilling system. To discover the COP_{OS} of the framework, all the components are interconnected as per the real system.

2.4 Economic modeling of the proposed hybrid energy system

An optimal mixture of a hybrid renewable energy-based absorption chilling system can be found to fulfill the cooling demand with maximum overall system performance (COP_{OS}) with the best economical feasibility. The economic methodology, based on the conception of capital cost, running cost, payback period, and net present value is determined to the finest level of the overall system economic analysis in this study.

The total aggregate expense is comprised by fittings and common works for the setting up of the plant. Government stipends are additionally reflected with the estimation of the capital expense. The parts considered in the plant for manipulating the capital expense of biomass, biogas and gobar gas plant, biogas and gobar gas burner and heat exchanger, and vapor absorption refrigeration system.

$$\begin{aligned}\text{Capital cost of HRES} = {} & \text{capital cost of BM, BG and GG plant} \\ & + \text{capital cost of VARS} \\ & + \text{capital cost of burner and heat exchanger in BG and GG plant} \\ & - \text{Govt.subsidies}\end{aligned} \tag{10}$$

The operating or running cost of a system comprises an operational and maintenance cost and yearly depreciation value.

$$\begin{aligned}\text{Running cost of HES} = {} & \text{cost of the energy resource used} \\ & + \text{operation and maintenance cost of energy conversion system} \\ & + \text{operation and maintenance cost of VARS} + \text{labour cost} \\ & + \text{depreciation value}\end{aligned} \tag{11}$$

The optimum economic value of a hybrid renewable energy-based refrigeration system includes the payback period to retrofitted with conventional cooling system with diesel genset, has been calculated as

$$Payback\ period = \frac{Incremental\ value\ of\ Capital\ cost}{Annual\ savings\ (Profit)} \tag{12}$$

where, the incremental value is the difference between the capital cost of fossil fuel-based cooling system and the hybrid renewable energy-based absorption refrigeration system.

$$\begin{aligned}&\text{Capital cost of fossil fuel} - \text{based refrigeration system} \\ &\quad = \text{cost of diesel generator} + \text{cost of fossil fuel} - \text{based cooling system}\end{aligned} \tag{13}$$

Annual savings is the difference between the running cost of diesel-based VCRS and the hybrid energy-based absorption cooling system. Where,

$$\text{Running cost of fossil fuel} - \text{based refrigeration system} = \text{cost of fossil fuel (diesel)} + \text{maintenance cost of refrigeration system} + \text{labor cost to operate the system} + \text{depreciation value} \tag{14}$$

Net present value (NPV) is the existing value of the costs of investment and running cost of a system over its period. Net present value of the investment is the difference between the present worth of the benefits and the costs ensuing from an investment.

$$NPV = \left[S.\left(\frac{(1+i)^n - 1}{i(1+i)^n}\right)\right] - CC \tag{15}$$

where 'S' stands for benefits at the end of the lifetime, CC is the initial capital investment of the system. The acceptance criteria of an investment project as: (a) NPV > 0, accept the system (b) NPV < 0, reject the system.

The life cycle cost (LCC) analysis based on present worth cost (PWC) method, which includes the initial capital cost, running cost, maintenance cost, replacement cost and salvage values is the useful prediction to analyze the various mixtures of hybrid renewable energy-based refrigeration system in remote locations. A life cycle of 18 years was used to evaluate the PWC of the system.

$$\text{LCC} = \text{initial capital cost} + \text{PWC}_{\text{RC}} + \text{PWC}_{\text{RE}} + \text{PWC}_{\text{SV}} \tag{16}$$

A forthcoming aggregate value for an item transformed into its corresponding existing value is called the present worth of this item. The following equation is used to calculate the PWC of operating costs and maintenance costs.

$$P_{RC} = RC.\left[\frac{(1+i)^n - 1}{i.(1+i)^n}\right] \tag{17}$$

The following equation is used to calculate the PWC of replacement costs and salvage values.

$$P_{RE} = RE.\left[\frac{1}{(1+i)^n}\right] \tag{18}$$

$$P_{SV} = SV.\left[\frac{1}{(1+i)^n}\right] \tag{19}$$

Here the salvage value of the hybrid energy-based absorption system after 18 years is estimated by assuming 7% of total initial costs of the system. The replacement cost of the hybrid energy-based absorption cooling system after 18 years is estimated by assuming 1% of total initial costs of the system.

3. Results and discussion

The results obtained from the modeling and simulation of the techno-economic performance of hybrid energy-based cooling systems for milk preservation in studied region was presented. In most of the studied regions the normal milk production is around 2500 L per day which can be cooled to the preservation temperature at 4°C in 4 hours. Therefore the 5TR system has been considered in this study. The performance parameters such as COP_{OS} (overall system COP), capital cost, running cost, payback period and net present value of a system are suitably analyzed in this chapter.

The amount of every energy source accessible, and the required quantity of energy resource to satisfy the cooling desires in an investigation area is appeared in **Figure 3**. It is seen that the accessible energy resources are less than the required amount. So any sole source accessible in this province does not satisfy the essential cooling demand. Along these lines, to meet the total cooling demand, all the accessible energy sources must be utilized in appropriate blends. Hence the identified three renewable energy resources, BM, BG and GG, are consolidated as BM-BG, BM-GG and BG-GG for this examination. In the graphs, the proportion of the two energy resources is plotted in X-axis so that the use of major energy source ascends from left to right though another energy source rises from right to left. Standard operational range represents that the base and most extreme availability of BM, BG and GG energy resources.

The overall performance of the cooling system with the ratio of energy sources is shown in **Figure 4**. It shows that, the overall system performance (COP_{OS}) increases with increase in the proportion of energy sources for the BM-GG and BM-BG mixtures, because of the high calorific rate of energy resources and the enhanced conversion efficacy of biomass gasifier. However, the COP_{OS} diminishes with increment in proportion of energy resources for the BG-GG mixture. In the mixture of BM-GG energy sources, COP_{OS} changes among 0.185 and 0.23. However the another two blends, BM-BG and BG-GG, indicates an moderately lowest COP_{OS} of 0.13–0.23 and 0.13–0.185 respectively. The BG-GG mixture indicates that the lowest value of COP_{OS}, so this mixture may not be considered in this area.

Figure 5 shows that the capital cost and running cost hybrid energy-based cooling system for different mixtures with the ratio of energy resources in the study area. It states that the BG-GG blends contribute the lowest capital cost, when the influence of the gobar gas contribution is at highest level. When the contribution of gobar gas source is higher than 70–80%, the capital cost variations of the mixtures BM-GG and BG-GG are under 10%. The variations of the capital expense of BM-BG illustrate a higher incentive than different blends. Hence, the BM-BG mixture is not appropriate in the study province. At the point when the impact of the gobar gas

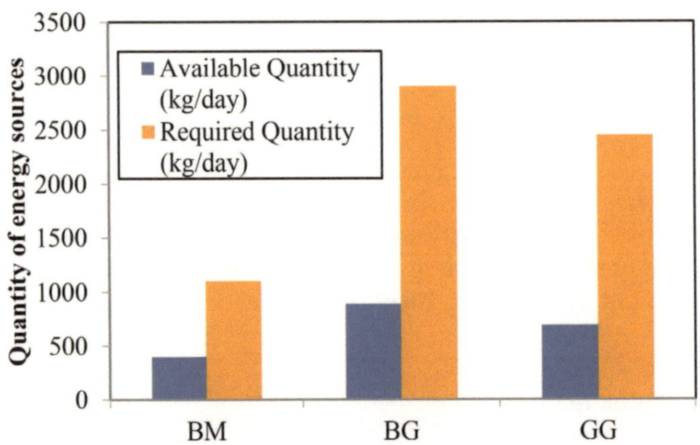

Figure 3.
Available and required quantities of energy sources in study region.

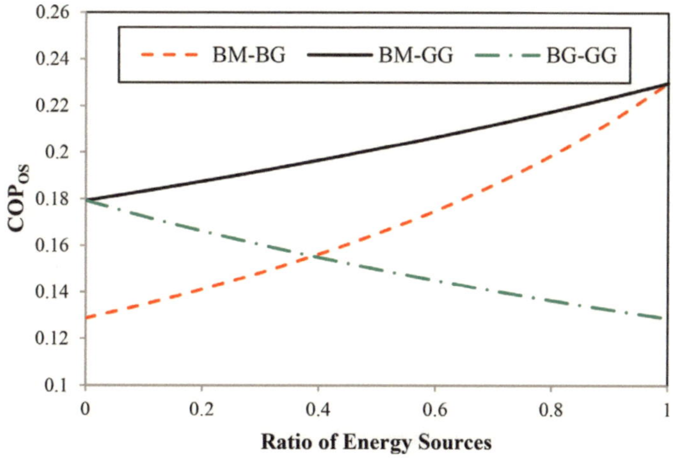

Figure 4.
Overall performances of cooling systems in the study region.

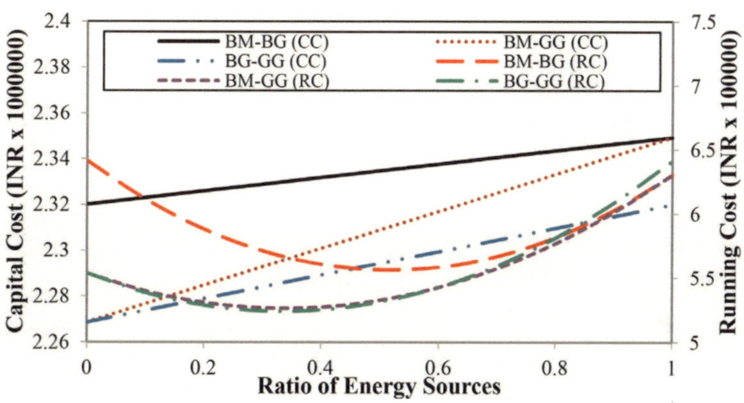

Figure 5.
Variations in capital and running cost of the hybrid energy-based cooling systems in the study region with various combinations.

commitment is 60–70%, the BM-GG and BG-GG blends demonstrate a low running expense contrasted with BM-BG blends. So, gobar gas influence is kept at 65–75% in BM-GG and BG-GG mixtures to attain a low running expense. So the mix of energy sources with gobar gas commitment is very essentialness to diminish the capital and running expense in the study province.

Figure 6 show that the variation of payback period of hybrid energy-based cooling system with the ratio of energy sources in study region. The mixtures of BM-GG and BG-GG illustrates a lowest payback period, when the influence of gobar gas contribution is higher level, as it reflects the lowest running cost of the process. It is identified that the overall system with the influence of 70% gobar gas contribution is suitable mixing value in that region. However, the influence of gobar gas contribution is more than 80% in the mixture of energy sources, a highest value of payback period has been observed. It concludes, it is not suitable to maintain the gobar gas contribution is above 80% in the mixture of energy sources. It is advisable that the influence of GG should be 60–70% in BM-GG and BG-GG mixtures to attain the lowest payback period.

Figure 7 shows that the net present value (NPV) of the various mixtures of hybrid renewable energy-based cooling system with the ratio of energy sources.

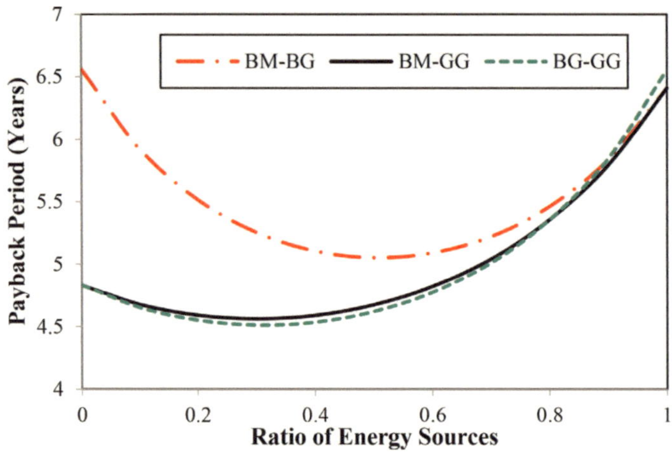

Figure 6.
Payback period of the cooling systems in the study region with various combinations.

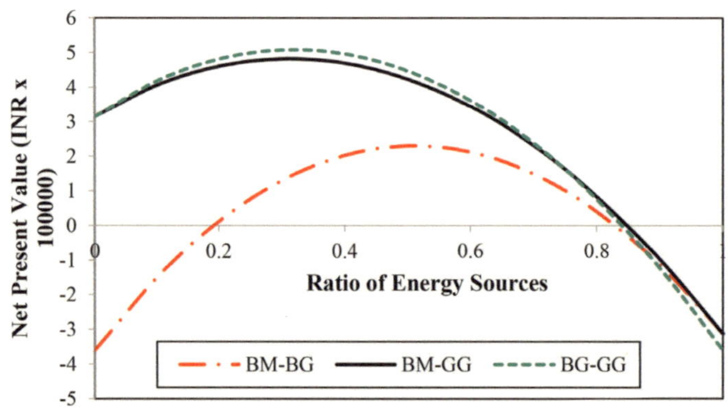

Figure 7.
NPV of the cooling systems in HR with various combinations.

When the influence of gobar gas is more than 20% in the ratio of energy mixture, BM-GG and BG-GG are shows the positive trends. Therefore these combinations are preferable once in the study region.

The life cycle cost (LCC) variation with the proportion of energy resources has been analyzed and noted in **Figure 8**. The pattern of BM-GG and BG-GG blends demonstrates that the LCC is low for BM-BG mixture, when the gobar gas commitment is kept at the most extreme. It is seen that the system with the GG influence of 70% is the most suitable meanwhile LCC worth is very low for this contribution.

The purpose of techno-economic analysis is to achieve, within a given system structure, a balance between the overall system performance and the various economical factors which will give an preeminent overall COP and cost of the system. In general, an overall system requires two conflicting objectives: one being an increase in overall system performance and the other is economical viability. Both of these factors should be satisfied simultaneously. The overall system performance is decided by the COP_{OS} and the economical viability is confirmed based on capital cost, running cost, payback period, net present value, life cycle cost, etc. Therefore the COP_{OS} and the economic parameters obtained from the study for the identified combinations are discussed in this section.

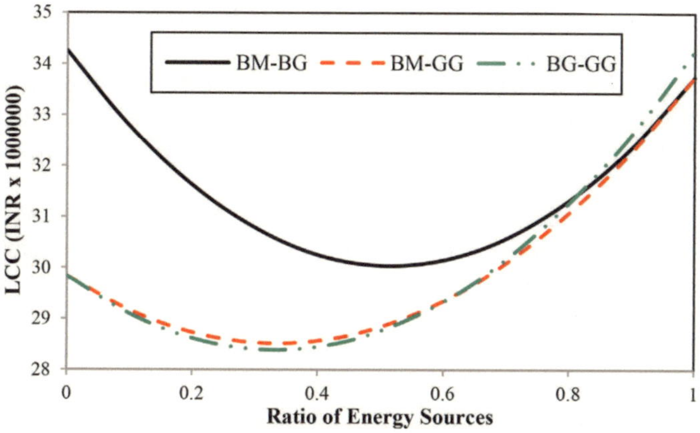

Figure 8.
LCC variation of hybrid energy-based cooling systems in the study region.

Sl. No	Combinations	COP_{OS}	CC (INR × 10^6)	RC (INR × 10^6)	PBP (years)	NPV (INR × 10^6)	LCC (INR × 10^7)
1	BM—BG	0.13–0.23	2.32–2.35	0.56–0.64	5.0–6.5	−0.36–0.22	3.0–3.42
2	BM—GG	0.18–0.23	2.27–2.35	0.53–0.63	4.6–6.4	−0.31–0.48	2.85–3.37
3	BG—GG	0.13–0.18	2.27–2.32	0.52–0.64	4.5–6.5	−0.36–0.50	2.84–3.42

Table 2.
Technical and economic performance of the hybrid energy-based cooling system in the study region.

The possible three combinations identified in the study region along with their COP_{OS} and economic factors are presented in **Table 2**. The values are obtained from **Figures 4–8**.

Table 2 demonstrates that the BM-GG mixture gives the most astounding COP_{OS} (0.185–0.235) because of the higher calorific range and conversion efficacies of biomass and gobar gas resources besides the lower running expense. Notwithstanding, this blend has higher capital expense than BM-BG and BG-GG mixture. This is because of the cost associated with the establishment of biomass plant. In addition, the payback period of BM-GG mixture is low and it's NPV and LCC is superior to the next two mixes. Hence, the hybrid energy system working with BM-GG mixture is the most suitable for the study province.

The simulation results expose that the gobar gas influence has extraordinary effect in all the economic values in the study provinces. Consequently 70% of gobar gas commitment in any mixture can be considered as the appropriate mixture in the study area.

4. Conclusions

The present investigation is focused on the development of environment friendly, energy efficient, 100% renewable energy-based cooling system for milk and agro preservation in remote areas. Alternative cooling systems have been studied by integrating the available renewable energy sources of biomass, biogas, and gobar gas with vapor absorption cooling system to solve the high energy demand and high emissions issues associated with conventional cooling systems. Based on the nature of activities taking place, the study area has been selected. The techno-economic study has been performed to identify the appropriate composition of available renewable energy sources to meet the cooling demand of a particular area. Based on the analysis the following conclusions are drawn.

In the study region, a single energy source cannot meet the energy required to produce the required cooling load. Therefore, a combination of energy sources could help us to implement 100% renewable energy-based cooling systems in villages.

In the study region, the combination of biomass and gobar gas (BM-GG) is the appropriate pair of energy sources. It offers higher COP_{OS} (0.18–0.23) due to the higher heating value and conversion efficiencies of biomass and gobar gas sources besides the lower running cost. Moreover, BM-GG combination requires lower payback period (4.6–6.4 years) in comparison with the other combinations such as BM-BG (5–6.5 years) and BG-GG (4.5–6.5 years). Additionally, the life cycle cost of the BM-GG combination is less than the other two combinations because of the lower running cost and replacement cost. Based on the reasons given above, the BM-GG combination-based cooling system can be considered as the most appropriate combination in the study Region.

All the observations show that the maximum usage of biomass energy source highly influences the COP_{OS}. Hence, BM could be utilized to the maximum in hybrid energy systems, to improve the overall performance.

The analysis demonstrates that the greatest utilization of biomass can improve the COP_{OS}, while the gobar gas diminishes the capital cost, running expense and payback period. Consequently, biomass and gobar gas resources could be used to the maximum in hybrid energy systems to enhance the COP_{OS}, and diminish the capital cost, running expense and payback period.

In this work, a new methodology has been developed to account for the sustainable cooling system to preserve milk and agro products for remote communities using renewable energy sources. The method applied here is generally applicable to any region.

Author details

Edwin Mohan[1*], Saranya Nair Mohan[2] and Joseph Sekhar Santhappan[3]

1 Department of Mechanical Engineering, University College of Engineering, Nagercoil, Anna University Constituent College, Nagercoil, India

2 School of Electronics Engineering, Vellore Institute of Technology, Chennai, India

3 Department of Mechanical Engineering, Shinas College of Technology, Sultanate of Oman

*Address all correspondence to: edwinme1980@gmail.com

References

[1] Kumar A. Solar-biomass hybrid cold storage-cum-power generation system for rural applications. Energetica India. 2012;**30**:54-55

[2] TIFAC. Agro-Food Processing Technology Vision 2020 Fruits & Vegetables—Current Status & Vision. Technology Vision 2020 Reports, TIFAC:V: 01: DR(FV); 2012

[3] Khan MJ, Iqbal MT. Dynamic modeling and simulation of a small wind-fuel cell hybrid energy system. Renewable Energy. 2005;**30**:421-439. DOI: 10.1016/j.renene.2004.05.013

[4] Khan MJ, Iqbal MT. Pre-feasibility study of stand-alone hybrid energy systems for applications in Newfoundland. Renewable Energy. 2005;**30**:835-854. DOI: 10.1016/j.renene.2004.09.001

[5] Khan MJ, Iqbal MT. Analysis of a small wind-hydrogen stand-alone hybrid energy system. Applied Energy. 2009;**86**:2429-2442. DOI: 10.1016/j.apenergy.2008.10.024

[6] Juhari AR, Kamaruzzaman S. Optimization of renewable energy hybrid system by minimizing excess capacity. International Journal of Energy Research. 2007;**1**(3):77-81

[7] Ashok S. Optimised model for community-based hybrid energy system. Renewable Energy. 2007;**32**: 1155-1164. DOI: 10.1016/j.renene.2006.04. 008

[8] Ashok S. Hybrid energy systems-prospects and initiatives at NIT Calicut. NITC-Research Review. 2009;**3**(1):3-16

[9] Singal SK, Varun SRP. Rural electrification of a remote island by renewable energy sources. Renewable Energy. 2007;**32**:2491-2501. DOI: 10.1016/j.renene.2006.12.013

[10] Rohit S, Subhes C. Off-grid electricity generation with renewable energy technologies in India: An application of HOMER. Renewable Energy. 2014;**62**:388-398. DOI: 10.1016/j.renene.2013.07.028

[11] Iniyan S, Sumathy K. An optimal renewable energy model for various end-uses. Energy. 2000;**25**:563-575. DOI: 10.1016/S0360-5442(99)00090-0

[12] Craig G, Eilin W, Kevin PM, Philip O. Regional integration of renewable energy systems in Ireland—The role of hybrid energy systems for small communities. Electrical Power and Energy Systems. 2013;**44**:713-720. DOI: 10.1016/j.ijepes.2012.08.012

[13] Rahman MM, Hasan MM, Paatero JV, Lahdelma R. Hybrid application of biogas and solar resources to fulfill household energy needs: A potentially viable option in rural areas of developing countries. Renewable Energy. 2014;**68**:35-45. DOI: 10.1016/j.renene.2014.01.030

[14] Alvares SG, Trepp TH. Simulation of a solar driven aqua-ammonia absorption refrigeration system part 1: Mathematical description and system optimization. International Journal of Refrigeration. 1987;**10**:40-48. DOI: 10.1016/0140-7007(87)90095-8

[15] Alvares SG, Trepp TH. Simulation of a solar driven aqua-ammonia absorption refrigeration system part 2: Viability for milk cooling at remote Brazilian dairy farms. International Journal of Refrigeration. 1987;**10**:70-76. DOI: 10.1016/0140-7007(87)90023-5

[16] Kaushika ND, Mishra A, Chakravarty MN. Thermal analysis of solar biomass hybrid co-generation plants. International Journal of Sustainable Energy. 2005;**24**(4):175-186. DOI: 10.1080/14786450500291909

[17] Prasartkaew B, Kumar S. A low carbon cooling system using renewable energy resources and technologies. Energy and Buildings. 2010;**42**: 1453-1462. DOI: 10.1016/j.enbuild.2010.03.015

[18] Prasartkaew B, Kumar S. Experimental study on the performance of a solar-biomass hybrid air-conditioning system. Renewable Energy. 2013;**57**:86-93. DOI: 10.1016/j.renene.2013.01.034

[19] Prasartkaew B. Mathematical modeling of an absorption chiller system energized by a hybrid thermal system: Model validation. Energy Procedia. 2013;**34**: 159-172. DOI: 10.1016/j.egypro.2013.06.744

[20] Prasartkaew B, Kumar S. The quasi-steady state performance of a solar-biomass hybrid cooling system. In: Proceedings of the Second TSME International Conference on Mechanical Engineering; 19-21 October 2011; Krabi. 2011. pp. 19-21

[21] Nixon JD, Dey PK, Davies PA. The feasibility of hybrid solar-biomass power plants in India. Energy. 2012;**56**: 541-554. DOI: 10.1016/j.energy.2012.07.058

[22] Jaishree P, Vijay VK. Experimental studies on drying of Zingiberofficinale, Curcuma longa l and Tinosporacordifolia in solar-biomass hybrid drier. Renewable Energy. 2005; **30**:2097-2109. DOI: 10.1016/j.renene.2005.02.007

[23] Iqbal MT. Modeling and control of a wind fuel cell hybrid energy system. Renewable Energy. 2003;**28**:223-237. DOI: 10.1016/S0960-1481(02)00016-2

[24] Jaramillo OA, Borja MA, Huacuz JM. Using hydropower to complement wind energy: A hybrid system to provide firm power. Renewable Energy. 2004;**29**:1887-1909. DOI: 10.1016/j.renene.2004.02.010

[25] Kaldellis JK. An integrated model for performance simulation of hybrid wind-diesel systems. Renewable Energy. 2007;**32**:1544-1564. DOI: 10.1016/j.renene.2006.07.004

[26] Li-qun L, Zhi-xin W. The development and application practice of wind-solar energy hybrid generation systems in China. Renewable and Sustainable Energy Reviews. 2009;**13**: 1504-1512. DOI: 10.1016/j.rser.2008.09.021

[27] Perez-Navarro A, Alfonso D, Alvarez F, Ibanez C, Sanchez I, Segura C. Hybrid biomass-wind power plant for reliable energy generation. Renewable Energy. 2010;**35**:1436-1443. DOI: 10.1016/j.renene.2009.12.018

[28] Fantidis JG, Mantzari VC, Kalkani E, Bandekas DV, Vordos N. A hybrid wind and hydroelectric power production system in Plaka, Alexandroupolis, Greece. International Journal of Advances in Engineering, Science and Technology. 2013;**2**(4): 376-385

[29] Lal DK, Dash BB, Akella AK. Optimization of PV/wind/micro-hydro/diesel hybrid power system in HOMER for the study area. International Journal on Electrical Engineering and Informatics. 2011;**3**(3):307-325

[30] Maherchandani JK, Agarwal C, Sahi M. Economic feasibility of hybrid biomass/PV/wind system for remote villages using HOMER. International Journal of Advanced Research in Electrical, Electronics and Instrumentation Engineering. 2012;**1**(2): 49-53

[31] Nahid-ur-Rahman C, Syed ER, Tofaeel AN, Mahabub A-A-F-I. Present scenario of renewable energy in Bangladesh and a proposed hybrid system to minimize power crisis in

remote areas. International Journal of Renewable Energy Research. 2012;**2**(2): 280-288

[32] Gupta A, Saini RP, Sharma MP. Steady-state modelling of hybrid energy system for off grid electrification of cluster of villages. Renewable Energy. 2010;**35**:520-535. DOI: 10.1016/j.renene.2009.06.014

[33] Gupta A, Saini RP, Sharma MP. Modelling of hybrid energy system-part I: Problem formulation and model development. Renewable Energy. 2011; **36**:459-465. DOI: 10.1016/j.renene.2010.06.035

[34] Gupta A, Saini RP, Sharma MP. Modelling of hybrid energy system-part II: Combined dispatch strategies and solution algorithm. Renewable Energy. 2011;**36**:466-473. DOI: 10.1016/j.renene.2009.04.035

[35] Gupta A, Saini RP, Sharma MP. Modelling of hybrid energy system-part III: Case study with simulation results. Renewable Energy. 2011;**36**:474-481. DOI: 10.1016/j.renene.2009.04.036

[36] Carrie H, Samuel G, Jonathan W. Evaluation of energy efficiency and renewable energy generation opportunities for small scale dairy farms: A case study in Prince Edward Island, Canada. Renewable Energy. 2014;**67**:20-29. DOI: 10.1016/j.renene.2013.11.040

[37] Zehetmeier M, Baudracco J, Hoffmann H, Hei Benhuber A. Does increasing milk yield per cow reduce greenhouse gas emissions? A system approach. Animal. 2012;**6**(1):154-166. DOI: 10.1017/S1751731111001467

[38] McSweeney PLH, Fox PF. Advanced Dairy Chemistry. New York: Springer; 2009

[39] Le Lostec B, Galanis N, Millette J. Experimental study of an ammonia-water absorption chiller. International Journal of Refrigeration. 2012;**35**: 2275-2286. DOI: 10.1016/j.ijrefrig.2012.05.012

[40] Koop K, Koper M, Bijsma R, Wonink S, Ouwens JD. Evaluation of Improvements in End-Conversion Efficiency for Bioenergy Production; Ecofys. 2010